Practical Reliability
Data Analysis for
Non-Reliability Engineers

For a complete listing of titles in the
Artech House Technology Management Series
turn to the back of this book.

Practical Reliability Data Analysis for Non-Reliability Engineers

Darcy Brooker
with
Mark Gerrand

ARTECH HOUSE

BOSTON | LONDON
artechhouse.com

Library of Congress Cataloging-in-Publication Data

A catalog record for this book is available from the U.S. Library of Congress.

British Library Cataloguing in Publication Data

A catalogue record for this book is available from the British Library.

Cover design by

ISBN 13: 978-1-63081-827-2

© 2021 ARTECH HOUSE
685 Canton Street
Norwood, MA 02062

Software accompanying this book can be found at:
https://us.artechhouse.com/Assets/downloads/brooker_827.zip

All rights reserved. Printed and bound in the United States of America. No part of this book may be reproduced or utilized in any form or by any means, electronic or mechanical, including photocopying, recording, or by any information storage and retrieval system, without permission in writing from the publisher.

All terms mentioned in this book that are known to be trademarks or service marks have been appropriately capitalized. Artech House cannot attest to the accuracy of this information. Use of a term in this book should not be regarded as affecting the validity of any trademark or service mark.

10 9 8 7 6 5 4 3 2 1

Contents

Preface   11

What This Book Is About   11
Intended Readers   12
Excel Versions   12
Excel Tools Included   13
Reliability Data   13
Acknowledgment   14
    References   15

Introduction   17

Red Flag: Software Reliability Analysis   19
Red Flag: Human Reliability Analysis   19
    References   20

1   Background: 10 Key Concepts Underlying Reliability Analysis   21

RAM   21
    Reliability   22
    Maintainability   23
    Availability   23
    Safety and Dependability   23
Probability and Statistics Terminology   24
    Probability   24
    Random   24
    Populations and Samples   24
    Parameters and Statistics   25
    Statistical Significance, Confidence Intervals, and Risk   25
    Valid Data   27
Data Types   27
    Censored Data   27

Discrete and Continuous	28
Independence	29

Appendix 1
Maintenance and Maintainability	31
Availability	32
Hidden Failures	33
Combining Event (e.g., Failure) Probabilities—Either Event Occuring	33
Complement Rule	33
General Additive Rule for Either of Two Events	34
Mutually Exclusive Events and Rare Event Approximation	34
Combining Event (e.g., Failure) Probabilities—Both Events Occuring	34
Conditional Probability	35
General Multiplication Rule for Both of Two Events	35
Independent Events	36
Bayes' Theorem	36

2 First Check Operating Conditions, Failure Modes, and Trends   37

With Nonrepairable Items, Check for Differing Operating Conditions and Failure Modes	38
Also Check Repairable Item Failure Data for Trends	39
How to Check Repairable Item Failure Data for Trends	41
Trend Analysis	43
GRP-EZ	46
Red Flag: Modeling Preventive Maintenance	46
Key Points	50

Appendix 2
IID, Stochastic Point Processes, Limit Theorems, and Repairable Items	51
IID	51
Stochastic Point Processes	51
ROCOF and Hazard Rate	52
Bathtub Curve for Repairable Items	52
Deterioration and Wear Out	52
Limit Theorems	53
Repairable Item Analysis Assumptions	53
Maintainers	53
Common Cause Failures	54
Intermittent Failures	54
Life Unit Selection	54
Replacement Parts	54
Proportional Intensity Models	54
Testing For Independence	55
Exponential Functions	56
Excel Least Squares Curve Fitting	57
Parameter Maximum Likelihood Estimates for The Power Law Repairable System Model	58
Maximum Likelihood Estimates of Parameters	58

Contents

Confidence Intervals on the Shape Parameter	59
Goodness-of-Fit	59
Tests For Linearity	60
Cost Models of Repairable Items	62
Minimal Repair Policies Based on Repair Cost Limits	62
An Age-Dependent Minimal Repair Policy for Deteriorating Repairable Systems	62
Adaptive Cost Policies	63
References	63

3 Nonparametric Data Analysis   65

Introduction	65
Sample Statistics	66
Mean	66
MTTF and MTBF	67
MTTR	67
Standard Deviation	67
Confidence Intervals and Nonparametric Analysis	68
Histograms	69
Distribution-EZ	70
Cumulative Plots	70
Example: Disk Drive Failures	71
Nonparametric Tests	72
Parametric Analysis Usually Follows Nonparametric Analysis	73

Appendix 3

Kaplan-Meier Method	75
References	77

4 Probability Distribution Representations   79

CDF, PDF, and Hazard Rate	79
CDF	80
PDF	81
Hazard Rate	82
Bathtub Curve for Nonrepairable Items	83
Common Reliability Distributions	86
Distribution-EZ	87
Sampling Distributions	87
Degrees of Freedom	89
Failure Modes	89

Appendix 4

Reliability of Electronic Systems	91
Integrated Circuits and Nonmicroelectronic Component Reliability	91
Key Stressors	91
Cables, Connectors, and Solder Joints	92
Intermittent Failures	92
Stresses and Product Life Acceleration	93

	Example: Data Center Cooling	94
	Chi-Square (χ^2) Goodness-of-Fit Test	95
	Kolmogorov-Smirnov (K-S) Goodness-of-Fit Test	96
	References	97

5 Weibull (Continuous) Distribution — 99

	Introduction	99
	Distribution-EZ	102
	Applying the Weibull Distribution Model	102

6 Life Data Analysis: Weibull Probability Plotting — 105

	Introduction	105
	Wei-EZ	107
	Choosing the Best Distribution and Assessing the Results	109
	Anomalies	113
	Exercise	116
	Further Reading	116
	References	116

7 Exponential (Continuous) Distribution — 119

	Memoryless Property	119
	Nonrepairable Items	120
	Repairable Items Caution	120
	MTTF	121
	Distribution-EZ	121
	Point and Interval Estimation for the Exponential Distribution Parameter	121
	Applying the Exponential Distribution Model	126
	References	127

8 Normal (Continuous) Distribution — 129

	Introduction	129
	Distribution-EZ	131
	Point Estimation of Normal Distribution Parameters	132
	Confidence Interval for the Mean	133
	Confidence Interval for the Standard Deviation	134
	Applying the Normal Distribution Model	135

Appendix 8
The Central Limit Theorem		139
	References	140

9   Lognormal (Continuous) Distribution   141

Charactertistics   141
Distribution Parameters   144
Parameter Estimation and Parameter Confidence Limits   145
Distribution-EZ   145
    Mean and Standard Deviation of Raw Data   146
    Applying the Lognormal Distribution Model   147
    References   149

10   Binomial (Discrete) Distribution   151

Introduction   151
Point and Interval Estimates for Binomial Distribution Parameter   152
Applying the Binomial Distribution Model   155

Appendix 10
Combinations   158
Acceptance (Pass/Fail) Testing   159
    References   161

11   Poisson (Discrete) Distribution   163

Introduction   163
Markov Models   167
Point and Interval Estimates for Poisson Distribution Parameter μ 168
Applying the Poisson Distribution Model   169
    References   170

12   Analyzing Degradation Data   171

Introduction   171
Nondestructive Degradation Sampling Using the Same Samples   172
    Example: Accelerometer Degradation   172
Destructive Degradation Sampling Utilizing a Common Distribution Parameter   174
Destructive Degradation Sampling, with No Common Distribution Parameter   176
    References   178

13   Preview of Advanced Techniques   179

Monte Carlo Simulation   180
Bayesian Analysis   181
Data Mining Methods   183
Prognostics and Health Management   183
    PHM Design   185
    PHM challenges   185
    Data-Driven PHM   186
    References   187

Bibliography 189

About the Authors 191

Index 193

Preface

What This Book Is About

Much of engineering focuses on designing and building products with improved performance—lighter, faster, and so on. But pursuing these objectives can make products less reliable. Accordingly, reliability problems are common.

Reliability engineering and management primarily focuses on preventing and minimizing failures and their consequences, whether those failures gravely threaten safety, reduce mission effectiveness, or simply increase costs. Perhaps this is why the U.S. Secretary of Defense Schlesinger said, "Reliability is, after all, engineering in its most practical form." Most engineering disciplines and systems can effectively use this body of knowledge.

Like other performance characteristics, reliability characteristics must generally be designed into a product—once the design is frozen, so is the product's susceptibility to failures from environmental stresses or wear processes. Further, rather than regarding reliability engineering as just applying a series of tools, the best outcomes are achieved when delivered through a systematic program and associated practices (and accompanying way of thinking). So, this book covers an important but small part of reliability engineering.

Nevertheless, statistical analysis of data is perhaps the area of reliability engineering that discourages the most people. Perfect information is seldom present in any decision-making situation however, and this is especially so in reliability engineering. Often, results or outcomes cannot be predicted in advance with certainty, even if essentially the same conditions exist (this might be likened, for example, to tossing a coin). In the face of uncertainty, probability (and statistical inference) acts as a substitute for complete knowledge. Probability and statistics theory allow us to generalize from the known to the unknown with high confidence.

This book presents basic probabilistic and statistical methods or tools used to extract the information from reliability data to make sound decisions. The tools presented can be applied across most technologies and industries, and the information and insights gained by doing so can help reduce whole-of-life costs, improve operating effectiveness, and improve safety.

Intended Readers

The intent of this book is to support people to get started to analyze reliability data, and to act as a refresher for those with more knowledge and experience. This book assumes readers have a level of mathematical knowledge at about grade 10 level or higher, covering exponential and logarithmic functions (including using such functions on calculators), as well as a basic proficiency in using Microsoft Excel ® spreadsheet (Excel) functions.

Optional supplementary material is provided in appendixes. However, any coverage of more advanced mathematics and references is relegated to notes and footnotes (for more advanced readers to pursue if desired, or for those interested in the fine print). You should read only the text without appendixes, notes, or footnotes on the first reading, and to then read more detail as the need (or interest) arises. This will make the reading much easier going.

For those readers who are not intending to become full time reliability engineers yet want to explore the reliability engineering field more broadly, we might suggest the book Practical Reliability Engineering by O'Conner and Kleyner [1]. Here you will find discussion that includes: specific failure mechanisms (e.g., fracture, fatigue, creep, corrosion, and vibration, as well as specific failure mechanisms for electronic systems); reliability and whole-of-life costs; specifying and contracting for reliability; designing for reliability; integrated reliability programs and organizational reliability capability; testing for reliability; accelerated testing; and reliability in manufacture and quality control; as well as common reliability engineering techniques (e.g., fault tree analysis (FTA), reliability block diagram (RBD) modeling, and Failure Modes, Effects, and Criticality Analysis (FMECA)).

Excel Versions

While a basic proficiency in using Excel spreadsheet functions is assumed, some Excel functions change with different releases and users may need to adapt some Excel functions shown to suit their own Excel version. If you are not sure exactly what function you need to use, begin by typing the function as shown in this book (including the = sign), and a list of related functions will appear. Selecting one by clicking on it will highlight the arguments of the selected function and clicking on the function name again will provide a full description of the function. Or use the Excel formula builder function *fx* in a similar way. The Excel Help tools can also assist.

Excel Tools Included

Some reliability analysis tools using Excel are also available with this book. These tools are not intended to replace commercially available data analysis software, such as available through Minitab and Reliasoft, but to support people to get started to analyze reliability data, and to transition to these much more capable (and costly) applications as needed.

Reliability Data

A final note by way of introduction: data analysis requires data. While the potential benefits of reliability data analysis have been outlined in broad terms, getting data might be considered the cost in any cost-benefit assessment. Reliability data can be relatively difficult to obtain, given it is acquired through observing and documenting failures, in use or through testing. Gathering such data can be time consuming—unless acceleration methods are used, this may span normal item lifetimes—and costly.

The focus of reliability testing also changes throughout the life cycle of a system or product and, with it, the type of data collected. While challenging to obtain, reliability data sets gathered during design and development are particularly valuable, given the much greater expense of corrections (redesign) later in the product life cycle. So, suppliers may perform tests on critical subsystems (such as engines) or components even before system prototypes are built. Often however, during conceptual design, and detailed design before the first prototype is built, comparisons to similar existing products (especially similar technology products used in similar ways and in similar environments), and a good deal of judgment must be used. When available, a first prototype can provide operating data that may be used to improve reliability, but the number of prototypes is not likely to be large enough to apply standard statistical techniques. However, engineers may apply a test-fix-test cycle and reliability growth testing to improve design reliability before any formal reliability tests are applied, and environmental stress testing may be used in conjunction with failure mode analysis, to refine the design reliability. Then, as the design is finalized and more product samples become available, life testing may be required for design verification. Later, during manufacturing, qualification and acceptance testing become important. Finally, reliability data collected throughout the operational life can contribute to finding and correcting defects that become apparent only with extensive field service, and also for optimizing maintenance (including parts replacement) schedules (as well as manufacturer warranty policies and improvements for the next model). At each of these phases, data sets are collected under widely different circumstances, which should be accounted for in subsequent analysis.

Because reliability data analysis often works with very limited data sets, all reliability data should be captured and used, not just failure data captured in a formal reliability program. Accordingly, reliability data often includes survival or right-censored data from systems that have not failed (yet)—the fact that an item has survived a particular time without failing is important information that should not be lost. The characteristics of small data sets and censored data require that reliability analysis uses some special statistical techniques (and assumptions based on engineering judgement). This book covers censored data and even ways to handle data with no failures at all, and provides ways for readers to assess a quantitative confidence range, which is perhaps more important for small data sets than a simple average (or mean).

(Hopefully) data is even scarcer for the safety engineer. When examining major historical accidents, such as the 1984 Bhopal chemical leak [2] and the 1986 Chernobyl nuclear reactor destruction [3], some of the difficulties of quantitative data analysis for safety engineering become apparent. For example, catastrophic events to be avoided may never have occurred before and have a very small probabilities because of redundant configurations of critical components. Reliability testing of the entire system may be impossible. Further, the catastrophic failures that have occurred were rarely the result of component failures only, or any single direct cause, but the result of a web of events. Arguably, this makes holistic safety analysis even more imperative, but safety analysis of hazards necessarily uses more qualitative methods and the safety engineer's broad understanding of potential hazards. Potential hazards are often identified by studying past accidents, using what-if techniques (such as FMECA), and paying close attention to field reports.

Acknowledgment

We wish to acknowledge Australian Defence Organisation reliability engineers, whose insight and experience inspired this book and tools.

References

[1] O'Conner, P., and A. Kleyner, *Practical Reliability Engineering*, Fifth Edition, John Wiley & Sons, 2011.

[2] Sriramachari, S., "The Bhopal Gas Tragedy: An Environmental Disaster," *Current Science*, Vol. 86, 2004, pp. 905–920.

[3] Shcherbak, Y., *Chernobyl*, New York: St. Martin's Press, 1991.

Introduction

One view of the nature of reliability data analysis is shown in simple form in Figure I.1. This book is concerned with phases 3, 4, and 5 in the figure. However, a full understanding of each phase comes only with experience. Moreover, data analysis is actually an iterative process (e.g., data may be subject to many analyses to gain insight).

The structure of the book is as follows. First, some underlying concepts are presented to provide a necessary basis for the later sections. Then (often missing or provided as an afterthought in many reliability textbooks) a chapter covers the need to check whether the data set reflects independent and identical distributions (IID), for nonrepairable items and especially for repairable items (and options to consider if they are not IID). Assuming IID can be reasonably assumed, the next part of the book outlines selecting the most appropriate probability distribution to represent the data, or whether no common distribution should be assumed and nonparametric analysis should only be performed. Assuming parametric models can be reasonably applied, the book then provides appropriate mathematical and Excel spreadsheet formulas to estimate parameters and confidence bounds (uncertainty) for the most common probability distributions used in reliability analysis. The final chapters provide an overview of analyzing degradation and introduce some more advanced reliability engineering techniques.

This book does not cover in any detail the planning and management of a reliability program, or with other reliability modeling or analysis techniques (apart from providing a foundation for understanding some of these). Further, this book focuses primarily on hardware and system reliability.

Indeed, while you are encouraged to seek reliability engineering professional help at any time you feel out of your depth, material that is significantly outside the scope of this book and where readers are encouraged to seek specialist reliability engineer support are identified in red flag para-

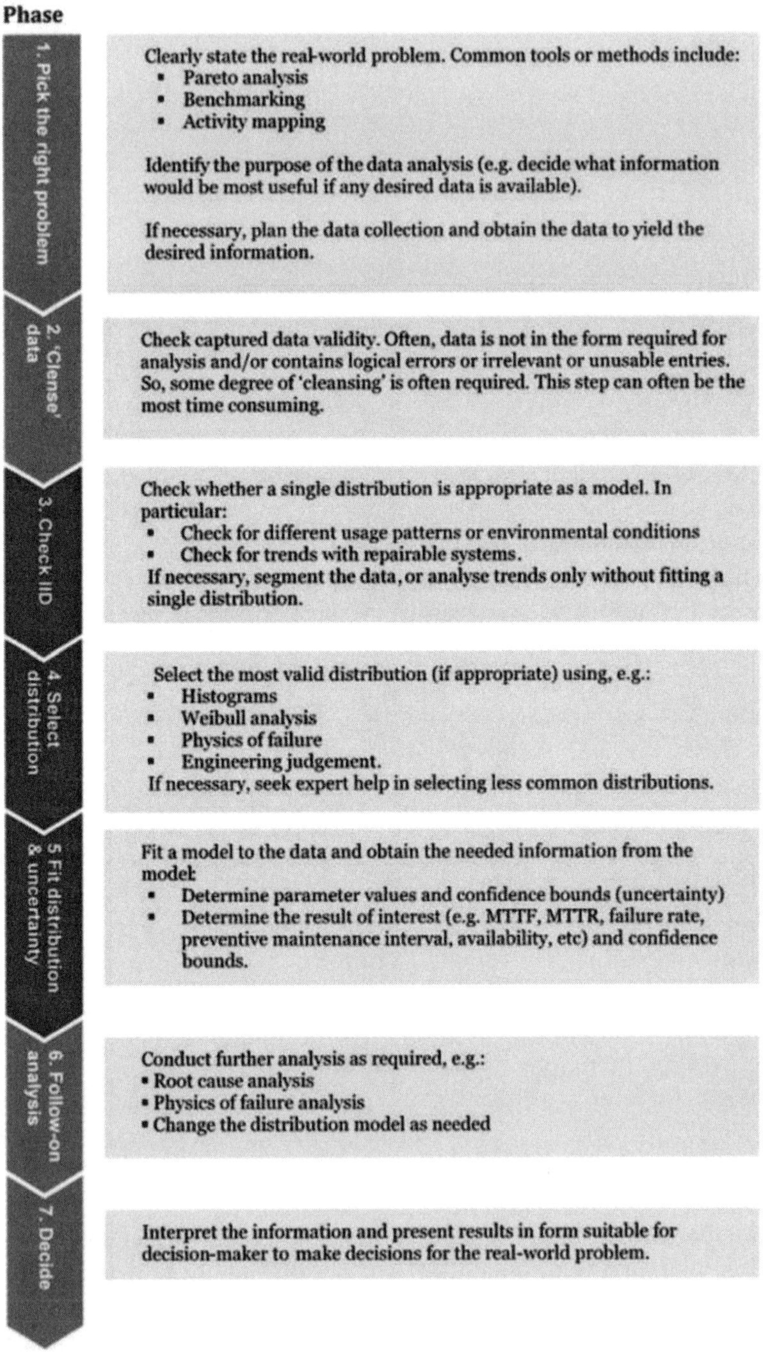

Phase	
1. Pick the right problem	Clearly state the real-world problem. Common tools or methods include: • Pareto analysis • Benchmarking • Activity mapping Identify the purpose of the data analysis (e.g. decide what information would be most useful if any desired data is available). If necessary, plan the data collection and obtain the data to yield the desired information.
2. 'Clense' data	Check captured data validity. Often, data is not in the form required for analysis and/or contains logical errors or irrelevant or unusable entries. So, some degree of 'cleansing' is often required. This step can often be the most time consuming.
3. Check IID	Check whether a single distribution is appropriate as a model. In particular: • Check for different usage patterns or environmental conditions • Check for trends with repairable systems. If necessary, segment the data, or analyse trends only without fitting a single distribution.
4. Select distribution	Select the most valid distribution (if appropriate) using, e.g.: • Histograms • Weibull analysis • Physics of failure • Engineering judgement. If necessary, seek expert help in selecting less common distributions.
5. Fit distribution & uncertainty	Fit a model to the data and obtain the needed information from the model: • Determine parameter values and confidence bounds (uncertainty) • Determine the result of interest (e.g. MTTF, MTTR, failure rate, preventive maintenance interval, availability, etc) and confidence bounds.
6. Follow-on analysis	Conduct further analysis as required, e.g.: • Root cause analysis • Physics of failure analysis • Change the distribution model as needed
7. Decide	Interpret the information and present results in form suitable for decision-maker to make decisions for the real-world problem.

Figure I.1 Phases of reliability data analysis.

graphs. The first two such red flags concern software reliability and human reliability.

Red Flag: Software Reliability Analysis

While hardware and software products can be managed throughout design and development in similar ways, analyzing software reliability differs from analyzing hardware and system reliability, because software differs from hardware in some significant ways. For example, while software does become obsolete, it does not wear out or deteriorate since it has no physical existence. And it is not manufactured but developed. Complications can also revolve around terminology such as the difference between errors, faults, and failures. Repairing or maintaining software is essentially modifying or redesigning the product (whether data or logic). Further, work to improve software reliability often focuses on programming style, languages, and software checking (including structured walkthroughs), as well as software testing, since any line of code and any item in a file can be a source of failure. Finally, while this book could be used to help analyze software reliability in a specific context, the results would not apply (in general) outside this specific context (or to similar functioning but different software in the same context) because of these issues, and others. Readers are advised to refer specific software reliability analysis tasks to specialists [1–3].

> **NOTE:** Practices in software engineering, and software reliability engineering, continue to evolve relatively quickly, but the books by Musa [1] and Lyu [2] are still used as references and IEEE has published a recommended practice for assessing and predicting software reliability [3]. This IEEE recommended practice suggests that, in the first instance, the following models should be considered for software reliability prediction: the Schneidewind model, the generalized Exponential model, and the Musa/Okumoto logarithmic Poisson model. Through its coverage of the Exponential and Poisson distributions, this book provides a solid basis to appreciate these software reliability models. The Software Engineering Institute of Carnegie Mellon University (http://sei.cmu.edu) is also a rich source of material relating to software engineering generally.

Red Flag: Human Reliability Analysis

Since human interactions with the environment and with fellow humans are extremely complex, and mostly psychological, much uncertainty surrounds how to probabilistically model human performance variability in real contexts. For example, when psychological stress is too low, humans become bored and make careless errors. When too high, people may make inappropriate, near-panic responses to emergency situations where clear procedures and thorough training is critical. Yet subtle combinations of malfunctions may still demand diagnostic and problem-solving ability beyond what can be achieved with procedural training. So, while studies have attempted to quantify various human error probabilities, the contexts of these studies are often very restrictive and we should exercise caution when using these models [7]. Again, readers are advised to refer specific human reliability analysis tasks to specialists.

> **NOTE:** Some qualitative approaches examine human error context by differentiating different types of causes of human error (e.g., Rasmussen's skills, rules, and knowledge errors [4]; or Reason's classifications of slips, lapses, mistakes, and violations [5]). Different causes may invoke different approaches for controlling human errors, such as better work organization, motivation, or training. More recent systems approaches shift emphasis to mechanisms and factors that shape human behavior. One systems approach to reduce the effects of human error is Reason's Swiss Cheese model [6] that focuses on conditions of work, trying to build defenses against human errors or mitigate their effects. The systems thinking approach regards human error as not a cause of accidents, but a symptom of the underlying system in which people work (e.g., see Dekker [7]).

References

[1] Musa, J. D., *Software Reliability Engineering: More Reliable Software Faster and Cheaper, 2nd Edition*, AuthorHouse, 2004.

[2] Lyu M. R. (ed.), *Handbook of Software Reliability Engineering*, IEEE Computer Society Press and McGraw-Hill, 1996.

[3] IEEE 1633-2016, *IEEE Recommended Practice on Software Reliability*, 2017.

[4] Rasmussen, J., "Skills, Rules, and Knowledge; Signal, Signs, and Symbols, and Other Distinctions in Human Performance Models," *IEEE Transactions on Systems, Man and Cybernetics*, Vol. SMC-13, No. 3, May-June 1983, pp. 257–266.

[5] Reason, J., *Understanding Adverse Events: Human Factors, BMJ Quality & Safety*, Vol. 4, No. 2, 1995, pp. 80–89.

[6] Reason, J., *Human Error*, Cambridge UK: Cambridge University Press, 1990.

[7] Dekker, S., *Ten Questions About Human Error: A New View of Human Factors and System Safety*, London: Laurence Erlbaum Associates, 2005.

CHAPTER 1

Contents

RAM

Probability and Statistics Terminology

Data Types

Independence

Appendix 1

Maintenance and Maintainability

Availability

Combining Event (Failure) Probabilities—Either Event Occuring

Combining Event (e.g., Failure) Probabilities—Both Events Occuring

Background: 10 Key Concepts Underlying Reliability Analysis

This chapter outlines 10 key background concepts underpinning the subsequent chapters:

- Reliability (and availability and maintainability) defined;
- Probability;
- Randomness;
- Populations and samples;
- Parameters and statistics;
- Confidence intervals and statistical risk;
- Data validity;
- Censored data;
- Discrete and continuous data;
- Independence.

RAM

The acronym RAM is short for the related concepts of reliability, availability, and maintainability.

21

Reliability

Reliability is commonly associated with successful operation, or no breakdowns or failures. However, for analysis, reliability needs to be defined quantitatively as a probability. Reliability is defined in four parts:

- The *probability* that
- An item will not fail (to perform an intended *function*)
- Under stated *conditions*
- For a specified period of *time*

All of these parts need to be identified and met for reliability to have any meaning:

- *Probability.* While reliability is a probability, other associated quantities can also describe reliability—the probability mean (mean time to failure (MTTF)) and failure rate are examples. With repairable systems, mean time *between* failure (MTBF) is also often used.

- *Functional failure.* When a system stops functioning totally—an engine stops running, a structure collapses, communications transmitter equipment provide no output signal—the system has clearly failed. However, in many instances, a failure in one circumstance might not be in another—a motor does not deliver a particular torque, a structure exceeds a particular deflection, or an amplifier falls below a particular gain. Intermittent operation or excessive drift in electronic equipment may also be defined as failures. Accordingly, failure needs to be defined quantitatively for these more subtle forms of failure, and a definition of failure often implies a required function so that we may determine when the equipment is no longer functioning properly.

- *Conditions.* Similarly, operating conditions (including the environment within which it must operate) determine the load or stresses the equipment is subjected to. Understanding loads or stresses is particularly important because the relationship between loads and failure is generally nonlinear—modest increases in relevant (failure producing) loads can dramatically increase failure probability (and associated failure rates).

- *Time (and other life units).* The way time is stated in reliability varies widely because the term is used generically to refer to some measure of *life units* for the item. Life units may actually be calendar time, operating time, kilometers, cycles, rounds fired, on-off switches, solid state drive erase cycles, and so forth. Selecting the most appropriate life unit is a very important part of specifying and analyzing reliability.

Maintainability

Maintainability relates to the ease of maintenance of an item. More formally, maintainability is:

- The *probability* that
- An item can be restored to a specified *condition*
- When maintenance is performed by personnel with a *specified skill level, using prescribed procedures and resources*
- In a specified period of *time*

The appendix discusses maintenance and maintainability in a little more detail, but maintainability analysis is not substantially considered further in this book, except to note that, given the parallels with our reliability definition, many of the techniques and models of reliability data analysis also apply to maintainability and maintenance data analysis.

Availability

Availability usually incorporates reliability and maintainability concepts. Availability is usually described either in terms of the probability that an item will function when required, or as the proportion of total time that the item is available for use, which might be defined in many ways (e.g., depending on the time frame of interest, instantaneous, mission length, planned service life) and what parts of downtime are reasonably in control or of concern. Except in the appendix, which discusses availability in a little more detail, availability data analysis is not considered further in this book.

Safety and Dependability

Sometimes the RAM acronym is extended to RAMS to include safety (or survivability). Safety addresses the potential for an item to harm people, other items, or the environment. And sometimes the RAM or RAMS acronyms are captured holistically with maintenance support performance and other attributes by the term dependability. These terms are not used further in this book.

Probability and Statistics Terminology

Supposedly identical products made and used (and often maintained) under identical conditions actually vary, in dimensions, in performance, and in life. Indeed, variability is inherent in all products. Such variability is described probabilistically or statistically.

Probability concepts are not generally intuitive. Some of the difficulty might be due to the language used, some might be due to an innate preference for certainty, and yet some of the difficulty is also because probability and statistics *is* difficult. Nevertheless, some fluency with basic concepts can be enormously helpful (not only for data analysis, but in daily life).

NOTE: Our preference for certainty may show in several ways. For example, we may prefer to interpret events as evidence confirming an extant worldview (thinking anecdotally) rather than as one of a range of events that have a certain probability of occurring—including some very low probability events (seemingly miracle events will occasionally happen). We may also tend to believe it when we see it, yet may not be able to see a physical average item.

Probability

The probability of an event (such as a failure) occurring is represented as a number between 0 and 1 (or between 0% and 100%) where 0 represents that an event (failure) can never happen, and 1 represents that an event (failure) is certain.

Random

The word random implies an equal or ascertainable chance, but not necessarily an equal probability. For example, if a bag contained seven red balls and three green balls only, *randomly* selecting a ball would give a probability of a red ball being selected as 0.7% or 70%.

Populations and Samples

A *population* is the collection of *all* items, or data under consideration. A *sample* is the part of the population from which information is collected. We analyze sample data to get information about populations. For example, when considering the average height of an Australian citizen, if 1,000 citizens are selected and their heights are measured, the population would be all Australian citizens and the sample would be the 1,000 selected. In many engineering (and business) problems, a population can be regarded as so large to be considered infinite and represented by a theoretical distribution.

Parameters and Statistics

A *parameter* is a descriptive measure of a population. A *statistic* is a descriptive measure of a sample that is used to provide estimates for the true population parameters (with associated uncertainty).

> **NOTE:** Most modern statistics uses the following notation:
> - Capital letters such as T or Y denote random variables—the thing we are measuring or counting—such as time to failure or number of maintenance actions;
> - Lower case letters such as t or $y_1 \ldots y_n$ denote observed outcomes or possible values;
> - Greek letters such as μ, σ, $\propto$, β, λ, and θ denote population or distribution parameters.

Different notation denotes statistics or parameter estimates. Latin letters can be used (to denote a statistic), or accents (e.g., hats or bars) can be used (e.g., to denote an estimate of the population parameter). So we might see the following for example:

	(Population) Parameter	(Sample) Statistic
Mean	μ	$\bar{x}$ or $\hat{\mu}$
Standard deviation	σ	s
Number of values	N	n

Statistical Significance, Confidence Intervals, and Risk

Like many specialists, statisticians have evolved their own terminology that can be confusing or misleading to nonspecialists. Statistical significance is one such term that, like confidence intervals, attempts to answer the question: how good is the (point) estimate of a population parameter, based on our sample data?[1] However, since a confidence interval is often easier for nonspecialists to interpret, this book focuses on confidence intervals. Nevertheless, we need to be clear about what confidence intervals and confidence limits mean.

Confidence intervals reveal the uncertainty of point estimates for parameter values.[2] Suppose that a sample is taken, and a parameter (such as

---

1. Statistical significance has its roots in ideas of hypothesis tests which indicate whether, given inherent random scatter in the data, a difference between a sample of data and a hypothesized model is small enough to be convincing. Hypothesis tests are closely related to confidence interval estimation and might roughly be considered the inverse.
2. Note that most data, and models, are inaccurate to some extent. Therefore, the uncertainty in any estimate or prediction is often greater than a confidence interval might suggest.

the average or mean) and a (say) 90% confidence interval of the parameter is calculated using the sample. We can say that the true (population) value of the parameter will be within the relevant confidence interval calculated for the sample 90% of the time if similar samples were to be hypothetically taken many times.

> **NOTE:** Perhaps because of the term confidence, confidence intervals are frequently misunderstood or misinterpreted as providing more certainty than is warranted. For example, an interval does not mean a particular probability of the population parameter lying within the interval (it either does or does not, there is no matter of probability), nor that a repeat experiment would provide the same probability of a parameter falling within the interval. Essentially, a confidence interval probability relates to the estimation procedure reliability, rather than the specific interval calculated.

We should not be surprised that, as our sample size increases (getting increasingly closer to the population size), the confidence in our point estimate also increases and our confidence interval will reduce (for a given confidence level). Using our average height of Australian citizens example, we should expect to have much greater confidence (relatively small confidence interval) in our estimate of the average if we used a random sample of 3,000 citizens, than if we used a random sample of 3.

Confidence interval calculations use the following notation and terms:

- The confidence level, denoted as $1-\alpha$, is the probability that the population parameter is in the given confidence interval.

- So α is the probability that the true population parameter *is not* in the confidence interval. We call α (alpha) the risk.

> **NOTE:** Referring back to significance terminology, this is also the statistical significance level. Readers should not confuse this idea of statistical significance with practical significance. Statistical significance attests that the result is not just random sampling variation or noise, whereas practical significance relates to the practical importance of any findings. So, while analysis results may be practically important, we should only rely on them when also statistically significant.

Common values of risk used are 1%, 5%, 10%, and 20%; the value selected depends on the application and the consequences of being wrong about our estimates. Note that a lower risk level (or higher confidence level) implies a wider confidence interval (with zero risk, or 100% confidence, that our parameter lies in the range $\pm\infty$).

We must also distinguish between one- and two-sided confidence levels. For example, a reliability and maintenance engineer would typically use:

- A one-sided lower confidence level on reliability (since this represents the conservative bound);
- A one-sided upper confidence level for maintainability (since this represents the conservative bound);
- Two-sided confidence levels on the parameters of the distribution.

Valid Data

Most statistical analysis assumes that sample data sets are randomly sampled from the population of interest.[3] A *random sample* is chosen by selecting units for a sample in such a manner that all combinations of items under consideration have an equal or ascertainable chance of being selected as the sample. Using the example of determining the average height of Australian citizens, samples from a subset of a population (e.g., professional basketball players, or one gender only) are *not* random samples and can give misleading information—as can samples from a different population (such as Chinese citizens). When faced with equivalent situations in reliability analysis (e.g., using laboratory test data to predict field failures) we must use engineering judgment to decide how well the data set represents the population of interest and how much weight to put on the information.

Data Types

Censored Data

Reliability data analysis is made more challenging because we often do not have precise or complete data, requiring special statistical methods. Four possible types of reliability data are:

- Exact times.[4]
- Right-censored data, in which we only know that the event (failure) happened or would have happened after a particular time (e.g., if an item is still functioning when a test is concluded). That the item sur-

---

[3]. Other sampling methods are sometimes used, including stratified sampling and two-stage sampling, and associated data analysis must take the sampling method into account. This book assumes simple random sampling.
[4]. Or other life unit.

vived up to a particular time is valuable information that should not be lost in analysis just because we do not have an exact failure time.

- Left-censored data, in which it is known only that the event (failure) happened before a particular time (e.g., if a failure is observed at the first periodic examination).
- Interval-censored data, in which it is known only that the event (failure) happened between two times (e.g., if the items are checked every five hours and an item was functioning at hour 145 but had failed sometime before hour 150). So, left-censored data can be considered a special case of interval-censored data.

To complicate the discussion on censoring further, slightly different analysis may be required for different test plans:

- Failure-censored data comes from tests that are stopped after a fixed number of failures (with time random);
- Time-censored data comes from fixed duration tests (with the number of failures random).

Discrete and Continuous

A random variable is discrete if its observed outcomes can be *counted*. Discrete random variables in reliability engineering are often binary—either present or not—such as defects (or other attribute), successful launches, or failures. If the observed outcomes cannot be counted but need to be *measured*, the random variable is continuous. Continuous random variables include height, time (and other life units) to fail, and maintenance time. Different probability distributions apply to continuous and discrete random variables.

> **NOTE:** Some problems can be recast from countable random variables to measurable and vice-versa. For example, while a random variable might be measurable, the problem can be recast as a discrete (binary) one by setting a (meaningful) threshold value. Further, while strictly countable, where large (countable) numbers are involved and small changes in counts are not meaningful (e.g., seconds or minutes when the time scale is days or weeks), the problem is often recast with the random variable as measurable.

For example:

- If employee records in an organization are viewed and the sex and marital status of each employee is noted, this would be discrete data;

- If a spool of cable is weighed and the weight recorded, this would be continuous data;
- If a spool of cable is weighed and the occurrence is recorded in one of a sequence of weight intervals, this would be discrete data;
- The time to failure of an aircraft engine would likely be continuous data;
- Whether the time to failure of an aircraft engine exceeds 10,000 operating hours is discrete data.

Independence

Determining a probability of a single event (such as a failure) invites questions of how probabilities of multiple different events are determined. While the appendix summarizes key probability rules,[5] for our purposes, we need to be clear about a common assumption made when analyzing data: independence.

Two events (e.g., failures) are called (statistically) independent if the events are not connected; for example, the probability of one event occurring does not depend on (the probability of) the other event occurring.[6] The independence assumption means that a component failure is not affected by any other component failing (or operating). When events are independent, our mathematics and modeling get much easier. Perhaps unsurprisingly then, such assumptions are often made and, sadly, often the chief justification for the independence assumption is its easier mathematics. Therefore, we should always keep in mind that such assumptions may not, in fact, hold making our results wrong (or at least indicating that we may need to modify our analysis or temper our results).

For example, subsequent rolls of dice or flips of a coin can reasonably be expected to be independent. But when we start considering real-life reliability and maintainability data analysis, the situation is seldom so clear cut. Failure processes can also introduce dependencies through loading, human error, and other common causes. Indeed, with repairable systems, independence of successive times between failures is not likely.[7] Assuming identically distributed lifetimes means assuming that the components are

---

5. Also, see probability textbooks for other Boolean Laws of Probability including the commutative law, associative law, and De Morgan's Theorem.
6. At least in ways that are not accounted for in our model.
7. People also often confuse independence and mutual exclusivity, but mutually exclusive events cannot be independent (except for the extreme case when $P(A) = 0$ or $P(B) = 0$) since the occurrence of one event precludes the other occurrence, and vice versa. (For mutual exclusivity, $P(A|B) = 0$, while for independence $P(A|B) = P(A)$.)

nominally identical, and exposed to approximately the same environmental and operational stresses, hence the need to consider some checks for independence and identically distributed data, the subject of the next chapter.

APPENDIX 1

Maintenance and Maintainability

Relatively few complex systems are designed to be totally maintenance-free. Maintenance can be generalized into two classes, preventive maintenance (PM) and corrective maintenance (CM).

PM aims to increase product reliability by staving off the effects of wear out mechanisms, such as wear, corrosion, and fatigue. With PM, parts are replaced, lubricants changed, adjustments made, and so on, before failure occurs. Monitoring methods are also used to provide indications of the condition of equipment and components. These might include nondestructive tests to detect fatigue cracks; temperature and vibration monitors on bearings, gears, or engines; and oil analysis to detect wear or breakup in lubrication and hydraulic systems.

CM actions return the system to service after failure. While the objective of PM is to increase reliability, most often the criteria used for judging CM effectiveness is availability, roughly defined as the probability that the system will be operational when needed.

Maintenance time includes several activity groups:

- Preparation time includes finding the person for the job, travel, obtaining tools, test equipment, publications, and so on;
- Active maintenance time is actually doing the job, which may include: diagnosis (especially for corrective maintenance where the failure cause is not obvious), studying maintenance manuals and so on, conducting actual hands-on maintenance, verifying that the maintenance is satisfactory, and post-maintenance paperwork when essential for the equipment to be released for use;
- Delay time (logistics and administration): waiting for parts and so on.

Human factors play a very strong role in maintenance, including training, vigilance, and judgement, as well as social and psychological factors that can vary widely.

In relatively complex repairable equipment, high reliability alone is not enough to ensure that it will be available when needed; it must also be maintained quickly (preferably while continuing to operate). However, ease of maintenance—maintainability—derives from a product's design. Maintainability designs incorporate built-in test equipment (BITE), modularity so that a failure can be corrected by easy replacement of the failed module, and accessibility for lubrication and calibration. Notwithstanding, active maintenance time is the only maintenance time (excluding documentation) that the designer can influence yet may prove to be only a small part of the overall maintenance time.

Availability

While reliability plays a central role in analyzing nonrepairable items, when considering maintenance (of repairable items), availability often becomes the focus of attention. Availability might be regarded broadly as the equivalent of reliability for repairable systems. Availability is sometimes defined as the proportion of total time that the item is available for use. The total time being considered depends on the context—it may be the time to accomplish a mission, or it may be the design life. The most general expression of availability is:

$$\frac{(\text{Uptime in required operating period})}{(\text{Uptime in required operating period}) + (\text{Downtime in required operating period})}$$

This expression highlights why different types/definitions of availability exist, because there are different answers to what is uptime (failure definition), what operating period are we concerned with, and what is included and excluded from downtime? Common variations of availability, used for different purposes are as follows:[8]

- *Inherent availability* (often used in design). The simplest case assumes no preventive maintenance (or else preventive maintenance performed when the item is not required), a constant rate of occurrence of failure (ROCOF) or MTBF and a constant mean repair rate (MTTR) or mean time to repair, no delays, and large operating period (where the instantaneous availability approaches the steady-state availability):

$$A_i = \frac{MTBF}{MTBF + MTTR}$$

- *Achieved availability* (often used in maintenance). Achieved availability can be considered the availability in an ideal support environment. Achieved availability excludes logistics and administrative downtime, but includes both preventive maintenance and corrective maintenance. Accordingly, instead of MTBF and MTTR, mean time between maintenance actions (MTBMA) and mean active maintenance time

---

8. Note that by applying definitions based on statistical variables necessarily means that the resulting availabilities are actually availability *estimates*.

(MAMT) or mean maintenance time (MMT) are sometimes used (assuming steady state availability):

$$A_a = \frac{MTBMA}{MTBMA + MAMT}$$

- *Operational availability* (of most interest to operators or users) measures actual average availability over a period of time, by including all sources of downtime (to determine a mean down time (MDT)). Accordingly:

$$A_o = \frac{\text{Average Uptime in needed time interval(s)}}{\text{Total needed time interval(s)}} \text{ or } A_o = \frac{MTBMA}{MTBMA + MDT}$$

Hidden Failures

An important class of failures, however, concerns unrevealed or hidden failures occurring (e.g., in systems that are not operated continuously, such as emergency or backup equipment, or stored repair parts that may deteriorate over time). Hidden failures (being in the standby mode) are not detected until an attempt is made to use the particular item and may be a primary cause of unavailability. A main defense against hidden failures is periodic testing, but this must be weighed against the costs of testing (and possible excessive wear from too-frequent testing). With some assumptions (asymptotic or longer term availability, constant failure rate λ, test interval $T \ll MTTF$, and test time t), the optimal test interval for hidden failures can be determined to be

$$T_{opt} = \sqrt{\frac{2t}{\lambda}}$$

Combining Event (e.g., Failure) Probabilities—Either Event Occuring

Venn diagrams are useful to visualize probabilities of either of two events occurring.

Complement Rule

Since the probability of an event either occurring or not occurring is certain, the probability that an event A will not occur, $\overline{A}$, is: 1 - (the probability that A does occur). See Figure 1.1.

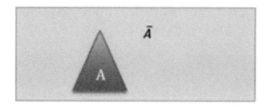

Figure 1.1 Complement rule.

General Additive Rule for Either of Two Events

In general, if events A and B can occur simultaneously (i.e., are not mutually exclusive), we must be careful to avoid double counting, so: P(A or B) = P(A) + P(B) - P(A & B). See Figure 1.2.

Mutually Exclusive Events and Rare Event Approximation

If events A and B cannot occur simultaneously, are mutually exclusive or P(A & B)=0, then the probability of A or B occurring is simply the sum of individual probabilities: P(A or B) = P(A) + P(B). When events are not mutually exclusive but are rare (i.e., P(A) and P(B) are both small) then the probability of events occurring together, P(A & B), can be very small. If P(A & B) is neglected, the rule for mutually exclusive events can be used as an approximation. See Figure 1.3.

Note that jargon differs between texts, for example:

Pr(A or B) = P(A+B) = A+B = P(AUB) = AUB.

Combining Event (e.g., Failure) Probabilities—Both Events Occuring

Contingency tables are useful to display two attributes of a population or sample and to calculate more complex probabilities. For example, suppose

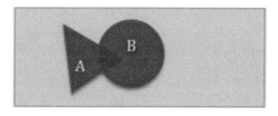

Figure 1.2 Either of two events occurring.

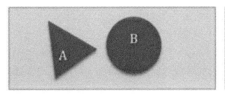 

Figure 1.3 Mutually exclusive events and rare event approximation to mutually exclusive events.

each part on a population of 192 balls can be one of four colors and one of three sizes according to a contingency table (Table 1.1):

Using such a table, the following calculations are straightforward: P(red) = 46/192 = 0.24; P(red or small) = P(red) + P(small) − P(red & small) = 46/192 + 70/192 − 16/192 = 0.521; and P(red & small) = 16/192 = 0.083.

Conditional Probability

Contingency tables are also useful to demonstrate conditional probability concepts. For example, what is the probability that a selected part is large, given (or on the condition) that it is known to be green: P(large, given it is green) = P(large | green) = 21/54 = 0.389. A key point is that this probability is generally not the same as the probability without conditions (e.g., P(large) = 65/192 = 0.338) since changing the initial conditions places a restriction on the events to be considered. It is useful to remember that the category to the right of the | in the conditional probability symbol points to the denominator in the probability fraction. A formal definition for conditional probability is: P(B|A) = P(A & B) / P(A).

General Multiplication Rule for Both of Two Events

Rearranging the conditional probability formula gives the general multiplication rule: P(A & B) = P(A) × P(B|A) or P(A & B) = P(B) × P(A|B).

Table 1.1 A Contingency Table

	Red	Yellow	Green	Blue	Total
Small	16	21	14	19	70
Medium	12	11	19	15	57
Large	18	12	21	14	65
Total	46	44	54	48	192

Independent Events

If P(A|B) = P(A), events A and B are called (statistically) independent (since P(A) does not depend on P(B), and vice versa). So, if events A and B are independent, then P(A & B) = P(A) x P(B).

Bayes' Theorem

Conditional probability concepts are the basis for Bayesian techniques. Using the conditional probability rule, and replacing the numerator using the general multiplication rule, we obtain statements of Bayes' theorem:

$$P(A|B) = \frac{P(A)P(B \mid A)}{P(B)} \quad \text{or} \quad R_{A|B} = \frac{R_A R_{B|A}}{R_B}$$

In this form, P(B) is called the *prior* probability of event B in the sense that it is the probability before knowing information about event A (this can be thought of as historical evidence). P(A|B) is called the *posterior* probability, given the prior evidence. The more general form permits a component to have what may be thought of as many states: Let $A_1 \ldots A_i$ be a set of events that are mutually exclusive and for which one member of the set must occur. Then

$$P(A_i|B) = \frac{P(A_i)P(B \mid A_i)}{\sum_{j=1}^{k} P(A_j)P(B \mid A_j)}$$

Note that jargon differs between texts, for example:

$$Pr(A \text{ or } B) = P(A + B) = A + B = P(A \cup B) = A \cup B$$

CHAPTER 2

Contents

With Nonrepairable Items, Check for Differing Operating Conditions and Failure Modes

Also Check Repairable Item Failure Data for Trends

How to Check Repairable Item Failure Data for Trends

Trend Analysis

Appendix 2

IID, Stochastic Point Processes, Limit Theorems, and Repairable Items

Proportional Intensity Models

Testing for Independence

Exponential Functions

Excel Least Squares Curve Fitting

Parameter Maximum Likelihood Estimates for the Power Law Repairable System Model

Tests for Linearity

Cost Models of Repairable Items

First Check Operating Conditions, Failure Modes, and Trends

Modeling and prediction can be improved if a known probability distribution or mathematical formula can reasonably approximate a set of data.[1] So, reliability engineering often seeks to determine a model of best fit to the data set and to derive associated parameters. Several factors should be considered when choosing the most appropriate model for a data set, including:

- Goodness-of-fit;[2]
- Past experience and engineering judgment;
- The nature of the data set.

In the rush to analyze an available data set, this last aspect can sometimes be overlooked. In particular, most common methods to derive distribution parameters only apply when data points are IID. If we incorrectly assume that the

---

1. This chapter primarily summarizes Ascher and Feingold [1].
2. Noting that the accuracy and credibility of any parameter estimations are highly dependent on the quality, accuracy, and completeness of the supplied data set.

data points are IID, we will obtain misleading results. Appendix 2 provides some additional background on IID and other topics relating to this chapter.

Challenges in reliability data analysis include few failure events and censoring (as discussed), but also aggregation. Aggregation particularly presents challenges when items are operated differently, in different operating environments, or maintained by different policies (or a combination of these), because doing so may change the pattern of failures from item to item. Accordingly, we need to categorize data into homogenous groups first.

So, the following principles should generally be first applied when confronted with a reliability data set:

- With nonrepairable items (commonly parts), check the data set for differing operating conditions and failure modes;
- With repairable items, also check the data set for age trends.

With Nonrepairable Items, Check for Differing Operating Conditions and Failure Modes

The identically distributed assumption may not hold when the items are operated in very different conditions (environments, operating characteristics, duty cycles, etc.). In such cases, we should resist the temptation to use a consolidated data set and, as much as possible, analyze data obtained under different conditions separately. Then, only if similar parameters for each are obtained should the data be consolidated (such as when the confidence bounds for parameters overlap significantly). Another, often related, indicator to separate our data set is if different failure modes are present (i.e., our items are failing in different ways).

Do not immediately treat the analysis as jumping into number crunching of the data set provided. Rather, consider the source of the data set, its context, and other information about the data set and, if appropriate, separate the data set into separate groups to analyze separately.

The key message is that we might make reasonably accurate predictions of failure when we can isolate key stressors, but if we were to consider all the data as a single set, our estimates necessarily mix information relating to multiple stressors, making our predictions much less accurate than they could potentially be (if we can make sense of the data set at all). More advanced techniques that attempt to determine the effect of different factors on reliability include proportional intensity models, outlined in the appendix.

Sometimes nonparametric analysis (such as histograms) and Weibull plotting (covered in later chapters) can also provide clues that data points are not identically distributed, but not always. Therefore, we should use all the information available to us to check that the IID assumption is reasonable. However, this check is particularly important for repairable systems, for reasons explained in the following sections.

Also Check Repairable Item Failure Data for Trends

Most real world complex products are not replaced after failure, but repaired. When considering repairable items, the IID assumption often does not hold so we should not attempt to apply a (single) probability distribution to repairable systems, except under very specific circumstances.

> **NOTE:** A nonrepairable item, often a component or part (ignoring one shot and go/no go items), is discarded and replaced the first time it fails. A repairable item, after failing to perform one or more of its functions satisfactorily, can be restored to perform satisfactorily by any method other than replacement of the entire item (including that no parts are replaced; repair might be made by an adjustment). Repairable items are often more complex systems.

Unfortunately, starting with this assumption tends to produce self-reinforcing but grossly invalid results, which can lead to inappropriate replacement decisions or preventive maintenance schemes. Therefore, repairable system reliability analysis is covered before looking at probability distributions.[3] While this section is based on complex-sounding stochastic (i.e., random) point process theory, the practical implications for basic analysis of repairable systems is straightforward.

> **NOTE:** When exact failure times are unknown or when repairable items are not operated over reasonably continuous time periods (e.g., one-shot devices such as missiles), different techniques, not addressed by this chapter, apply. Nevertheless, many of the chapter's sentiments apply also to these situations. Readers should also note that, while many concepts in this chapter can also apply to reliability growth during system development, this chapter assumes that no design or configuration changes occur to change the inherent reliability of the repairable item.

Essentially, we should not ignore the basic concept that if a repairable item fails more frequently with increasing operating time, its reliability deteriorates (and if it fails less frequently with increasing operating time, its reliability improves). For example, consider the problem of choosing a preferred set of failure data in Figure 2.1 [1]. While we can easily distinguish

---
3. The principles outlined also directly apply to maintenance analysis.

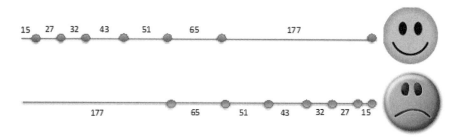

Figure 2.1 The order of failures often matters with repairable systems.

between these data sets, if we ignore the order of occurrence, both data sets appear the same.

So we can resolve the bulk of this problem (of assuming identically distributed data points) quickly and easily, by first checking for trends in data collected chronologically.[4] If no increasing or decreasing trend exists, the rate of change of failures (ROCOF) is relatively constant, suggesting that the repairable item is repaired to a same as new condition (i.e., a renewal process exists[5] and the distribution models discussed in later chapters may be used).

However, a large class of repairable items do not exhibit renewal behavior (i.e., constant ROCOF). In these cases, times between failures (interarrival times) cannot be assumed to be IID; the interarrival times are not coming from the same distribution and cannot be fitted to a single probability distribution (such as those discussed in later chapters). Accordingly, if age analysis shows a decreasing or increasing trend with time, then interarrival times should not be analyzed separately.

If a reparable item is not being renewed, two possibilities are as follows:

- If an item retains the same wear characteristics it held before it failed and underwent repair, it has undergone minimal repair and its condition is same as old.[6] We commonly accept same as old as a first-order model for a car. Often the first thing a used car purchaser wants to know is the year of manufacture or odometer reading. If used car

---

4. The interarrival times of a repairable item are always naturally ordered chronologically, so they should be analyzed similarly. Indeed, if any data set appears in any type of meaningful order, chronological or otherwise, the data should be initially analyzed in that order.
5. In which case the process is called stationary and is modeled by an ordinary renewal process.
6. This situation is modeled by a nonhomogeneous Poisson process (NHPP). A NHPP describes situations where the probability of a given number of failures in a given time period is a Poisson process whose parameter is a function of time.

purchasers assumed renewal, then they would ask for the time or distance travelled since the last servicing or repair.

- The other possibility is that the condition after repair is better than old, but worse than new.[7]

If we omit this step when analyzing repairable items, that is if we assume that the data points are IID, we might order the data points in rank order and then analyze the rank ordered data set to show an apparently constant failure rate (exponential) distribution. This result is clearly misleading when there is an increasing or decreasing ROCOF.

Therefore, when analyzing repairable items, we need to first consider reliability growth or deterioration as a separate phenomenon (distinct from part burn-in or wear out, outlined in a later chapter).

NOTE: The confusion and mistakes made between repairable item deterioration/growth and part wear out analysis may be partly attributed to the utility of the power law formula similar to the Weibull hazard rate formula to describe each. A key difference however is that the random variable is not interarrival times (times between failure or repair) when performing repairable system deterioration/growth analysis, but system age. Note also that, when describing system deterioration/ growth, this formula does not relate to a distribution: there is no requirement for any area under a PDF curve, or CDF curve, to stop at 100%—since renewal can occur the population is growing. Similarly, maintenance times need to be analyzed separately for age trends first, before attempting to fit any distribution.

How to Check Repairable Item Failure Data for Trends

For chronologically ordered times between successive repairable item failures, the order of analyzing data matters:

- Do not pool repairable item failure data (at least initially). If data for two or more repairable items exists, pool test *results* from the individual repairable items rather than pool the data themselves.

- Check for trends. Test successive times between failures for trend. A trend exists if there is a tendency for successive times between successive failure to decrease (corresponding to deterioration) or increase (corresponding to improvement).[8] A trend shows that a common

---

7. This situation is modeled by a generalized renewal process.
8. This loose definition excludes cyclic trends.

distribution of successive times between failures does not exist, so investigating properties of such a nonexistent distribution would be meaningless. If a trend exists in the successive times between failures, then renewal is not occurring and we should base subsequent analysis on the chronological trend only, using models outlined in the next couple of pages. Note that reordering trending data points by magnitude tends to make the data set appear to have a constant failure rate.

- (Ideally) check for independence. If no trend exists, the times between successive failures have identical (marginal) distributions but are not necessarily independent. Ideally, we should next test the successive times between failures for independence. In general, estimators based on incorrectly assuming independence will give greater sampling error than when independence holds but, in sharp contrast to the trending case, at least the properties of a common distribution are being estimated. Therefore, this step is often omitted (see the appendix for testing for independence).

- Check for renewal processes. If we find no trend (nor statistical dependence), for example, the number of failures (and hence repairs) is proportional to the system age, then we might reasonably conclude (since no evidence exists to the contrary) that renewal is occurring and that the inter-arrival times are from the same (IID) distribution, such as one of those commonly used for nonrepairable failure time analysis.[9] In this case, we should consider the most appropriate distribution as well as distribution-free (nonparametric) techniques. This step does involve reordering the IID times between failures by magnitude (i.e., forming order statistics). This is the subject of the remaining chapters of this book.

NOTE: Some reliability guidance documents claim that if no trend is indicated, analyses based on the exponential distribution may be validly used. However, a linear plot on linear paper does not imply an homogeneous Poisson process; indeed, any renewal process will yield an approximately straight line, as will any stationary process (i.e., a process whose parameters do not change over time). The justification for the guidance is that, using limit theorems (i.e., in the long run) the only important parameter of the renewal process is the mean interarrival time, so an homogeneous Poisson process and associated exponential distribution (being the simplest to model) will be an adequate approximation to the underlying renewal process. However, this is too sweeping a statement

---

9. Formally, lack of trend is not sufficient for stationarity; it is possible that the mean ROCOF might remain constant, even though the detailed structure of the process changes. However, in practice, this possibility is usually ignored.

because we may still be concerned about whether wear out is also taking place. In particular, we should distinguish between a homogeneous Poisson process and a renewal process whose times between successive failure indicate rapid wear out.

Trend Analysis

Repairable system deterioration/growth analysis is relatively straightforward: in essence, plot system age (rather than interarrival times) against cumulative failure, and check whether the pattern is reasonably linear. We can use some simple graphical tools to do this. If we plot cumulative failures versus cumulative time on linear axes, the successive times between failure of a deteriorating system tend to become smaller, so the plot will tend to be concave up (equivalent comments apply for improving systems).

> **NOTE:** Since linear plots will tend to mask local variation because of the (monotonically) increasing nature of the plot, even for relatively large sample sizes, Ascher and Feingold [1] advise that if doing this by hand consider dividing the total observation interval into three or more equally sized subintervals (depending on the number of failures) and plot the estimated average ROCOF for each subinterval. A deteriorating system will show a tendency for successive estimates of ROCOF to increase (equivalent comments apply for improving systems). Regardless of the choice of subinterval length, this procedure usually does indicate trends.

The expected number of failures in these situations are commonly modeled as a power function (because of their minimal repair basis) where the expected number of failures M is given by:

$$M(t) = at^b$$

where a is a scale parameter, b is a shape parameter, and time t is cumulative time. (See the appendix for a refresher on exponential functions.) Estimation of these parameters using Excel can be achieved by plotting an XY scatter graph of system age vs. failure number, then fitting a power trend line to the graph and displaying the equation on the chart. The parameter estimates can then be read directly from the power curve equation.[10]

---

10. Note, however, that this calculation does not account for any additional time during which no failure occurred (i.e., it assumes the data set is failure censored).

NOTE: The formula $M(t) = at^b$ refers to the Crow-AMSAA model [2], an extension of the Duane model [3]. Duane introduced the technique of plotting this function on log-log paper. A linear plot, subject to sampling variability, indicates a good fit to the function. Further, the slope of the plot estimates $(1-b)$. Note, however, that plotting on log-log paper will tend to straighten any function, so a linear plot on such paper does not necessarily indicate that the most appropriate model is this function.

This power law model has the same form as the hazard rate function $h(t)$ for the Weibull distribution, but the times referred to are different—system age and interarrival times respectively. The power law model makes no assumption that the system age times are distributed as an IID random variable. This potentially misleading mistake is not helped with the power law model sometimes referred to as a Weibull process in the literature.

The instantaneous MTBF of this model as a function of time is given by:

$$MTBF_t = \frac{t^{1-b}}{a}$$

Other forms of this model exist (e.g., in terms of ROCOF m(t) rather than number of failures:

$$m_i(t) = \alpha t^{-\beta}$$

where

$$\alpha = ab \text{ and } \beta = 1 - b.$$

Here, β is the reliability growth slope of the Duane plot. (In terms of the parameters for M(t),

$$a = \frac{\alpha}{1-\beta} \text{ and } b = 1 - \beta).$$

Among other models are the Cox and Lewis log-linear model, virtual age models, and proportional (covariate) hazard models.

Applying this to our sad face data set in Figure 2.1 shows Excel-calculated estimates for b and a of 2.26 and 8×10^{-6} respectively, as shown in Figure 2.2.

Most importantly, this graph shows that times between successive failures are not identically distributed. Therefore, assuming the system is same-as-new after repair and applying a single lifetime distribution model makes no sense. We *can* say that the system is deteriorating. For example, (using the dashed Excel trend line parameter values) we can extrapolate

Trend Analysis

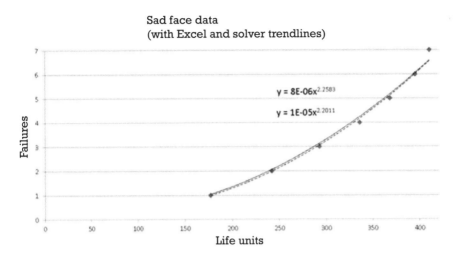

Figure 2.2 Applying the Excel power law model to sad face data set.

the eighth and ninth failure times to T=452 and 475, much smaller times between successive failure than the time to first failure.

With a little more effort, we can calculate a better least squares estimate for the parameters (b and a of 2.20 and 1×10^{-5} respectively). See the appendix.

We might also calculate parameters based on a maximum likelihood estimate (MLE) for b in this example (b and a of 3.42 and 8.2×10^{-9} respectively) and a 90% confidence interval for the parameter ($1.87 \leq b \leq 6.75$). See the appendix.

NOTE: Statistical analysis packages often estimate a parameter of a distribution, given sample data points, using a maximum likelihood estimate (MLE) method. When a distribution type is known but a parameter of the distribution (Y) is unknown, the PDF can be written f(x, Y) where x represents the values of the elements of the distribution. If $x_1 .. x_n$ is a random sample from the distribution, we can define a likelihood function L(Y) as the probability that these n values will be the ones selected, so L(Y) = f(x_1, Y) f(x_2, Y)...f(x_n, Y), and find the value of Y that will maximize L(Y). This is done by setting its derivative to zero and solving the resulting equation. (The logarithm of the likelihood function is often used as it is more tractable and gives the same result.) This process gives the value of Y that maximizes the likelihood that the randomly selected numbers will have values $x_1 .. x_n$.

We can also check linearity of repairable system age vs cumulative failures more formally than the eyeball test (see the appendix).

We can use the information from our trend models to develop better cost models for repairable systems (see the appendix).

GRP-EZ

Alternatively, we can use the Excel General Renewal Process (GRP) analysis tool provided, GRP-EZ, to perform similar calculations numerically for us, if we select the minimal repair (same as old) option.

GRP-EZ also performs analysis of good as new and better than old, but worse than new assumptions, as well as providing graphical representations. GRP-EZ analyzes the better than old, but worse than new case based on the work of Kijima [5]. Kijima used the idea of virtual age through a degree of repair or repair effectiveness factor, Q. (Q = 1 corresponding to good as new perfect repair or renewal, and Q = 0 corresponding to 'same as old', minimal repair.) Kijima model I, whose realism is often questioned, assumes that repairs only address deterioration and damage created in the last period of operation. Kijima model II assumes that repairs address all deterioration and damage accumulated up to the current time. By its nature, we might reasonably regard CM or repairs as being closer to minimal repair than renewal.

Red Flag: Modeling Preventive Maintenance

While the GRP-EZ tool does not model PM, we might expect different repair effectiveness factors for CM and PM, with PM acting as a censoring of the failure (& CM) process. Readers are advised to refer combined CM and PM modeling tasks to specialists.[11]

Example: Aircraft Air Conditioner Failure Trend Analysis

Times to failure (in hours) of an air conditioning system installed in a single aircraft are shown in Table 2.1 [6]:

Table 2.1
Aircraft Air Conditioning Failure Data Set

90	100	160	346	407	456	470	494	550	570
649	733	777	836	865	983	1008	1164	1474	1550
1576	1620	1643	1705	1835	2043	2113	2214	2422	

---

11. Proportional intensity models may be used to address combining CM and PM repair effectiveness, as well as more sophisticated Arithmetic Reduction of Age (ARA) or Arithmetic Reduction of Intensity (ARI) models.

Trend Analysis

Plotting an XY scatter graph of system age vs. failure number then fitting a power trend line to the graph and displaying the equation on the chart using Excel allows parameter estimates to be read directly from the graph. See Figure 2.3, which also includes better least squares parameter estimates outlined in the appendix (b and a of 0.807 and 0.0565 respectively using the solid trend line). GRP-EZ using the minimal repair option provides b and a values of 0.9007 and 0.026 respectively.

This analysis indicates a slightly decreasing failure trend but is close to being linear. Some reliability engineers would regard this as close enough to renewal and go on to analyze times between failures (interarrival times), as described in later chapters.

Example: Gas Pipeline Compressor Failure Trend Analysis

Consider the failure data set at Table 2.2 for a repairable gas pipeline compressor system (between 2012 and 2014 [7]):

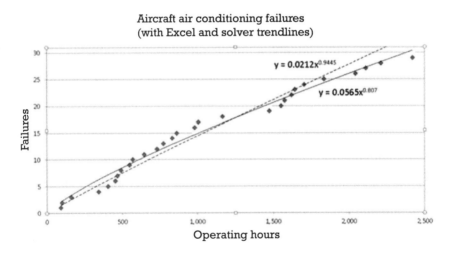

Figure 2.3 Aircraft air conditioning failures.

Table 2.2
Gas Pipeline Compressor Failure Data Set

Operating time (h)	576	1437	1979	2349	2685	3348	4200	4625	5147	5973	6546
Cumulative failures	14	37	46	50	56	62	70	72	77	86	91

Even though actual failure times are missing in this data set (they are interval censored), we can still plot these data points, though we would usually assume that the failures occurred in the midpoint of each interval instead of at the end and make the appropriate adjustments.

Again, we plot an XY scatter graph of system age vs. failure number then fit a power trend line to the graph and display the equation on the chart using Excel, allowing parameter estimates to be read directly from the graph. See Figure 2.4, which also includes better least squares parameter estimates outlined in the appendix (b and a of 0.7404 and 0.1541 respectively using the solid trend line). GRP-EZ using the minimal repair option provides b and a values of 0.7791 and 0.1002 respectively.

Figure 2.4 shows that the compressor system reliability is increasing (the curve is clearly convex) and we can quickly predict this trend using the parameters derived. Reliability engineers would be less likely to regard Figure 2.4 as close enough to renewal and so would simply go on to make judgements based on the trend parameters obtained.

Example: Submarine Diesel Engine Failure Trend Analysis

The final example in this chapter considers a failure data set for a single diesel engine on a submarine. See Table 2.3 [1].

Figure 2.5 shows our XY scatter graph and power trend line fit using Excel, as well as better least squares parameter estimates outlined in the

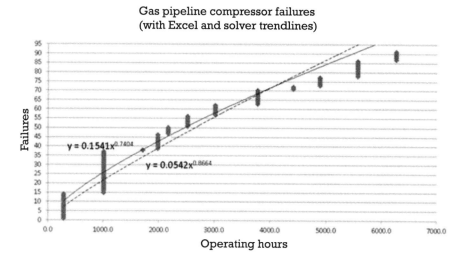

Figure 2.4 Gas pipeline compressor failures.

Trend Analysis

Table 2.3
Submarine Diesel Engine Failure Data Set (Thousand Operating Hours)

1.382	9.794	18.122	20.121	21.378	21.815	22.311	23.305	25.000
2.990	10.849	19.067	20.132	21.391	21.820	22.634	23.491	25.010
4.124	11.993	19.172	20.431	21.456	21.822	22.635	23.526	25.048
6.827	12.300	19.299	20.525	21.461	21.888	22.669	23.774	25.268
7.472	15.413	19.360	21.076	21.603	21.930	22.691	23.791	25.400
7.567	16.497	19.686	21.061	21.658	21.943	22.846	23.822	25.500
8.845	17.352	19.940	21.309	21.688	21.946	22.947	24.006	25.518
9.450	17.632	19.944	21.310	21.750	22.181	23.149	24.286	

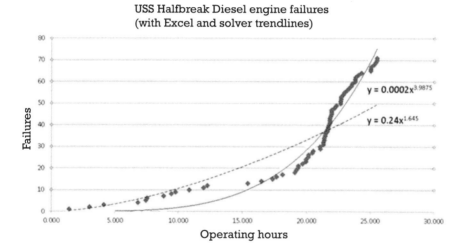

Figure 2.5 Submarine diesel engine failures.

appendix 2E (b and a of 3.9875 and 0.0002 respectively using the solid trend line). GRP-EZ using the minimal repair option provides b and a values of 2.7604 and 0.0093 respectively.

Figure 2.5 shows that the diesel engine reliability is decreasing rapidly, for example, failures are occurring more frequently (despite seven overhauls, not shown, at 9.453, 11.528, 11.933, 15.058, 17.315, 24.006, and 25.000 thousand operating hours). Attempting to fit a distribution to the times between failures from this data set would be a gross mistake since renewal is clearly not occurring from repairs. We should simply go on to make judgements based on the trend parameters obtained. (Some readers might also rightly question whether some change occurred in the engine usage or environmental conditions at around 18,000 operating hours, but no further information is available.)

Key Points

The take-home message is that the IID assumption is often not appropriate to model repairable systems. While such an assumption can be powerful and useful in initial analysis and modeling of complex repairable systems, we should test this assumption where possible (e.g., where actual data exists). And analyzing repairable systems is straightforward:

- We should pool test results from the individual data sets rather than pooling the data themselves (at least initially).

- We must consider the chronological order of times (or other life units) between successive repairable item failures until/unless we show that chronological order does not matter. Only if a set of successive times between failures do not exhibit trend can we reasonably represent them by a renewal process and apply similar techniques as used to analyze nonrepairable items.

APPENDIX 2

IID, Stochastic Point Processes, Limit Theorems, and Repairable Items

IID

We often assume IID data observations.

Identically distributed means that each observation is drawn from a fixed (stationary) probabilistic model. For example, consider rolling dice. Each of the six values always occurs with equal probability (i.e., rolling a die has a fixed probabilistic model) or each observation is identically distributed. An example of data observations that are not identically distributed is taking samples (especially large samples) without replacement, since the population changes with each sample.

Independence means that each observation does not depend on any other observation. Again, the outcome from rolling a die does not depend on any other roll. Independence also implies that the probability of observing two values can be calculated as the probability of observing the first multiplied by the probability of observing the second. So, if we throw two dice together, the values they display are IID random variables. If we throw the dice separately, the values they display are IID random variables. Note, however, that IID implies an element in the sequence is independent of what came before it, not that the probabilities for all elements must be the same—for example, repeated throws of loaded dice will still produce a sequence that is IID, despite the outcomes being biased. An example of data observations that are not independent is waiting times for each person (or item) in a queue, since the wait time of a person will depend, at least in part, on the wait time for the person in front.

Stochastic Point Processes

Many repairable item failures are not IID—failures are often not samples from any single distribution (and dependencies between failures often exist). Further, assuming that repairable items are just collections of unrepairable items is often wrong. Instead, repairable systems are modelled by stochastic point processes.

In the context of reliability, a stochastic point process mathematically models repairable item failures distributed randomly in time (roughly speaking, since repair times are disregarded). We call the rate of change (over time) of an expected number of repairable system failures the ROCOF (assuming that the expected value of number of failures, is absolutely continuous).

ROCOF and Hazard Rate

Do not confuse ROCOF with the hazard rate of nonrepairable items outlined in Chapter 4. The interpretation of ROCOF is the probability that a failure (not necessarily the first) occurs in a specified time interval. Contrast this with hazard rate as the conditional probability of (first and only) failure at a time, *given survival to that time*. No matter how complicated the situation becomes, when estimating the hazard rate, the normalizing factor is the reciprocal of the number of parts that have survived to that age. When the ROCOF is estimated however, there is no need to normalize by the number of survivors since all repairable systems are survivors! Expressed differently, the hazard rate of a probability distribution is a relative rate (relative to the number of survivors) whereas the ROCOF of a point process is an absolute rate. These rates cannot be equivalent.

Bathtub Curve for Repairable Items

Since the ROCOF of a repairable item can also change over its life, a bathtub curve concept for a repairable item can also be useful. A bathtub curve is characterized by an initially decreasing rate, a period of a relatively constant rate, followed by an increasing rate. Therefore, two bathtub curves exist, with key differences in interpretation. Chief among them is that the bathtub curve for a repairable item can be interpreted for the failure pattern of a single repairable item—in contrast to nonrepairable items where the bathtub curve represents a distribution of failure times for a population. The human mortality (deaths) rate curve is often discussed as an analogy when discussing nonrepairable item hazard rate. But we often ignore the existence of a second bathtub curve for humans, representing the pattern of illnesses which an individual (who attains old age) undergoes, essentially a ROCOF curve. A person undergoes reliability improvement and deterioration while people also undergo infant mortality and wear out. We must keep track of whether we are talking about illnesses or death.

Deterioration and Wear Out

Similarly, deterioration is not equivalent to wear out. The interpretation of the bathtub curve for a repairable item in the long-life region is that failures—of a single system—tend to occur more frequently (i.e., the system is deteriorating). However, increasing hazard rate of components need not imply deterioration of the repairable system. A repairable system consisting solely of nonrepairable parts that wear out (through various physical processes including wear, corrosion, fatigue, etc.) will not necessarily deteriorate since the system will be partially renewed if the original parts wear out and fail at around the same time. Indeed, system reliability growth can occur with its parts wearing out within each operating interval. Moreover, limit theorems indicate that the system's failure pattern will eventually settle down to a constant ROCOF even if all its constituent parts wear out. This distinction is important notwithstanding that, since a point process consists of a sequence of interarrival times, each of which has a part failure distribution associated with it, point processes and distributions are intimately related.

Limit Theorems

Limit theorems indicate that a repairable (series) system will eventually settle down to an homogenous Poisson process (HPP), a special case of renewal that is essentially a sequence of exponentially distributed IID times between successive failures (i.e., interarrival times). However, concluding that the HPP should universally model repairable systems is incorrect. Studies looking at series systems under relatively realistic conditions (e.g., a finite number of component parts, replaced one or more times, in finite periods) show highly nonstationary behavior for long time periods, particularly if their originally installed parts wear out rapidly at around the same time [8]. This invites the obvious question about the time required to approximate the steady state (the HPP). It takes a repairable system a long time to reach steady state because many or most parts must have been replaced at least once to assure a random mix of part ages, leading to an HPP. But many practical systems do not last long enough to reach the steady state.

Many different tests and estimators rely on asymptotic properties, but this is predominantly because of mathematical ease—we should test this assumption when we have data available (i.e., we should confront models with data). If we could always rely on asymptotic results and/or the HPP, why do we overhaul or replace systems because of reliability problems?

Repairable Item Analysis Assumptions

Readers should also note that many factors are ignored in modeling repairable systems using stochastic point processes. For example, most models do not consider the following factors.

Maintainers

In many maintenance operations the most important reliability factor may be the expertise of the maintainer. The Reactor Safety Study [9] showed that human error was one of the largest causes of malfunctions in nuclear reactors. In contrast, particularly effective repairs may improve the system, and improved operating and service manuals as well as operator and maintainer learning curves may have a strong positive effect. During repair, poor but unfailed parts may also be replaced, invalidating renewal process modeling.

Common Cause Failures

System failures may damage or cause failure of other parts, and two (or more) parts may be susceptible to the same environmental stress.

Intermittent Failures

Drift failures and part instabilities can cause reliability effects.

Life Unit Selection

Life units related to on/off cycling and environmental effects may be more important than system operating time. Kujawski and Rypka [10] investigated the effects of on-off cycling on several naval systems and found values as high as about 10:1 were reported for the ratio of ROCOF under cycling to ROCOF under continuous operation. Moreover, these stresses may change over time, for example, through seasonal effects.

Replacement Parts

Replacement parts may not be from the same population as the original part.

Proportional Intensity Models

In some situations, even though any individual repairable item has not failed often, a large number of similar items are available for analysis. A common approach has been to assume that times to first failure of these items are independent samples from one distribution, as are times between the ith and $(i+1)$st failures. However, when a large number of item copies are available for analysis there usually are known differences between them, which may have a marked effect on their reliability—the same type of item may be operated on different platforms, in different positions on a given platform (e.g., a multiengine aircraft), with different stresses, operators, and maintenance personnel, and so on. And different items may have different configurations.

In the past, the two main ways of handling these differences have been (l) to ignore them, or (2) to break the data set into two or more groups based on major differences. Even the second approach has shortcomings however, since compartmentalizing the data set into too many groups, may not allow meaningful analysis of any one group, and ignores the fact that the systems are very similar. Therefore, what is needed is a regression analysis that provides a common baseline for all systems, yet allows for the presence of censored data and uncertainty about suitable failure distributions.

So-called proportional hazard or proportional intensity models help to meet this need, including the Prentice-Williams-Peterson (PWP) model [11–12], which is an extension of Cox's model [13]. Using multiple regression techniques, the PWP model enables the analyst to consider such factors as different environmental stresses and the effects of different operators. In general, the entire history of each similar item can be considered to determine the effect of different factors on reliability. (However, computer programs are needed to perform the calculations.)

Particularly if insufficient data or numbers of similar copies are available for analysis, qualitative analysis techniques, such as root cause analysis (RCA) and failure modes effects analysis (FMEA), may also be used to investigate some of the factors outlined.

Testing for Independence

When no evidence of trend exists, we can consider successive times between failure of a system to be identically distributed, but not necessarily independent. For example, times might be positively correlated (e.g., large values might tend to follow large ones, or small ones might tend to follow small ones). Therefore, we should (desirably) next test for independence of successive times between failures.

The most straightforward way of testing for dependency of successive times between successive failures is through using the concept of lag. A lag is a fixed time displacement and lag plots should not exhibit any identifiable structure with random (independent) data points. While lag plots can be generated for any lag values, the most commonly used lag is 1, and a plot of lag 1 is a essentially a scatter plot of interarrival times along the horizontal axis vs the corresponding next interarrival time along the y axis. Most statistical software programs, however, use autocorrelation plots, for data values at varying lags, for checking randomness (independence)—autocorrelations should be near zero for any and all lag separations if the data points are independent.

However, in general, independence of successive times between failures is not likely. Except under special circumstances (such as all parts in series and with constant hazard rates), we should expect dependence between successive times to failure since, at the completion of the repair of the first failure, the age of the unreplaced parts is equal to the time to first failure.

For the usual case of a system composed of many parts, most repairs replace only a very small percentage of its constituent parts (possibly none). With the minimal nature of most repairs, the age of most of the parts in the system will be the sum of all the times to failure. We might also expect dependence because the same environmental stress may affect two or more parts.

However, even if dependence is found, we can still meaningfully estimate the properties of their common marginal distribution. In general, estimators incorrectly based on the assumption of independence will have greater sampling error than when independence holds but, in sharp contrast to the (nonstationarity) case of a trend existing, at least we estimate the properties of a common distribution. Accordingly, this step is desirable, but not always essential. Further, the lack of discussion of testing for independence in reliability engineering literature might be due to the need for a large number of successive times between failures (interarrival times), at least dozens and preferably 100 or more.

Exponential Functions

Recall that exponential functions are of the general form $y = a^x$ and can be used to describe a wide range of growth and decay processes (e.g., in chemical reactions, population growth, nuclear chain reactions, positive feedback, and economic growth or decay). Logarithmic functions are the inverse of exponential functions: $x = log^a\ y$ and logarithmic scales are also used widely, especially where the base (a) is 10 (e.g., decibel scales for levels of sound, pH scales for levels of acidity, f-stop levels for photographic exposures, and the Richter scale for seismic wave levels).

Another particularly important base is the number e, an irrational number of approximately 2.718, giving the natural exponential form $y = e^x$ and $x = \ln y$. This number, e, might best be understood by considering a compound interest example. Suppose that we had a sum of money invested for one year, with a (nominal) annual interest rate of 100%, accruing once per year. At the end of the year, the value of the investment would be 2 × the initial sum. If the interest accrued twice per year, at the first half-year the investment would be valued at (1 + 50% x 1) = 1.5 × the initial sum, and at the end of the year, the investment value would be 1.5 × this amount = 2.25 × the initial sum. As the accrual period reduces (and the number of accrual periods increase), the value of the initial investment keeps rising, but at a tapering (asymptotic) rate according to the equation:

$$Final\ value = initial\ value \left(1 + \frac{1}{n}\right)^n$$

where *n* is the number of accrual intervals in the period of interest (e.g., year). As *n* keeps increasing (towards infinity), the final value asymptotes towards 2.718 (= *e*). This example illustrates that the number *e* is associated with continuous growth (or decay) processes. Indeed, *e* can be defined as

$$\lim_{n \to \infty} \left(1 + \frac{1}{n}\right)^n$$

as well as many other ways.

Excel Least Squares Curve Fitting

In Excel, you can create an XY (scatter) chart and quickly add a best-fit trendline based on various functions, as well as obtain a goodness-of-fit (R^2) indication.

However, the approach that Excel uses to fit (nonlinear) trendline curves means that the fitted values do not truly minimize sum of squared deviations and that the reported value for R^2 is incorrect. This is because Excel's method first transforms the Y-axis so that a perfect trendline would form a straight line, then Excel applies standard linear regression to these transformed values.

For example, when fitting an exponential curve $y = a \times e^{bx}$ to a data set, Excel will take the logarithm of both sides of the exponential formula, which then can be written as $\ln(y) = \ln(a) + bx$. With $\ln(y)$ as the dependent variable and *x* as the explanatory variable, Excel then finds the intercept and slope that minimize the sum of squared deviations between actual $\ln(y)$ and predicted $\ln(y)$. Accordingly, the goodness-of-fit (R^2) value also refers to the transformed line. In practice, this approach tends to weigh data points unequally, causing differences when compared with direct least squares curve fitting.

Sometimes the Excel trendline approach is good enough, but where we need more accurate fits, or where we need to apply trendlines not covered, or distributions, to data sets, we can use Excel Solver to obtain actual least squares fits as follows.

- First, install the Excel Solver Add-In. Use Excel's Help tool if you are unsure how to do this; just search for Solver.

- Then set up an Excel spreadsheet with our data points and the curve formula that we wish to fit, similar to the Figure 2.6:

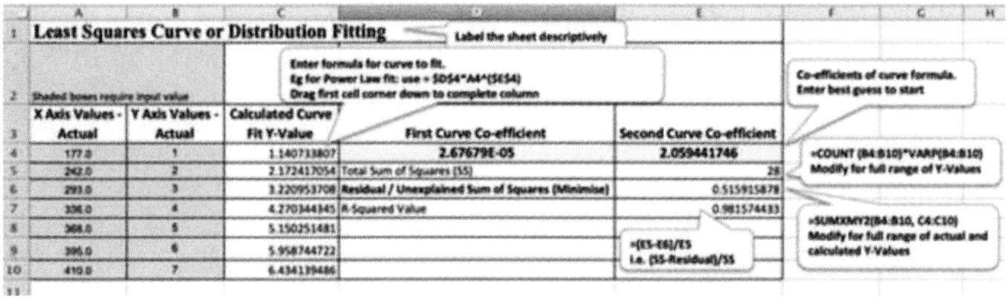

[Figure 2.6 table image]

Figure 2.6 Least squares curve fitting setup.

- Finally, apply Solver by selecting Tools/Solver, and set the objective as minimizing the residual sum of squares cell in the Excel spreadsheet, by changing the cells containing the coefficients of the formulas for the curve or distribution we wish to fit. Solver will attempt to change the coefficient values such that the residual sum of squares is minimized. See Figure 2.7.

Parameter Maximum Likelihood Estimates for Power Law Repairable System Model

The power law model form of NHPP models has been commonly applied to repairable systems [14]. Here, the expected number of failures, M, at cumulative time t is at^b, where $a, b > 0$ and $t \geq 0$.

Maximum Likelihood Estimates of Parameters

Assuming time-terminated tests the maximum likelihood estimates (MLE) of the parameters of this model are:

$$\hat{b} = \frac{m}{\sum_{i=1}^{m} \ln\left(\frac{T}{T_i}\right)} \quad \text{and} \quad \hat{a} = \frac{m}{T^{\hat{b}}}$$

where T_i represents the ith failure (arrival) time for m failures and T is the total test time. (When data points come from multiple copies of the product, this calculation needs to be modified [15].)

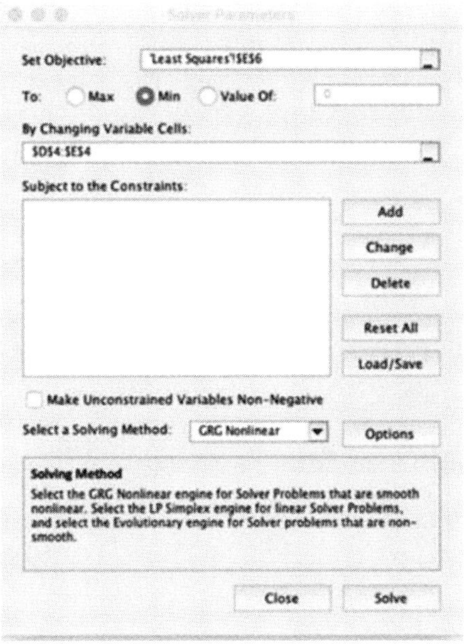

Parameter Maximum Likelihood Estimates

Figure 2.7 Least squares curve fitting solver.

Confidence Intervals on the Shape Parameter

For a two-sided 90% confidence interval on the shape parameter b in the model presented (also known as Crow Bounds), first calculate $\hat{b}$. Then calculate D_L and D_U as follows (e.g., using the χ^2 (Chi-squared) distribution):

$$D_L = \frac{\chi^2_{0.05}(2m)}{2(m-1)} \quad \text{and} \quad D_U = \frac{\chi^2_{0.95}(2m)}{2(m-1)}$$

The lower two-sided 90% (or lower one-sided 95%) confidence limit on b is $D_L \hat{b}$ and the upper two-sided 90% (or upper one-sided 95%) confidence limit on b is $D_U \hat{b}$.

Goodness-of-Fit

A goodness-of-fit test for this model is the Cramer-von Mises statistic given by:

$$C^2(m) = \frac{1}{12m} + \sum_{i=1}^{m}\left[\left(\frac{T_i}{T_m}\right)^{\hat{b}} - \frac{2i-1}{2m}\right]^2$$

where m is the number of relevant failures and $T_1 < T_2 < \ldots T_m$.

Critical values for this statistic at the 90% confidence level are provided at Table 2.4. If the $C^2(m)$ statistic exceeds the critical value corresponding to m in the table, then the hypothesis that this power law model adequately fits the data set should be rejected; otherwise the model should not be rejected.

Tests for Linearity

Linearity tests include the following (noting that more than five failures are usually necessary):

- The *Crow-AMSAA Test* suggests that we can confidently say that the trend is not linear using a Chi-square test statistic if

$$\frac{2r}{1-\hat{b}} \text{ is } < \chi^2_{2r, 1-\alpha/2} \text{ or } > \chi^2_{2r, \alpha/2}$$

  where r is the number of failures.

- The *Laplace Test* (also known as the Centroid Test) distinguishes between an HPP and a (monotonic) trend. Where the first m failure (arrival) times (not times between successive failures or interarrival times) are $T_1, T_2, \ldots T_m$, and T is the total time under test, we calculate the following, which quickly approximates a standardized normal variable (i.e., z-score):

Table 2.4
Cramer-von Mises Statistic Critical Values at 90% Confidence Level

m	Critical value	m	Critical value	m	Critical value	m	Critical value
3	0.154	8	0.165	13	0.169	18	0.171
4	0.155	9	0.167	14	0.169	19	0.171
5	0.160	10	0.167	15	0.169	20	0.172
6	0.162	11	0.169	16	0.171	30	0.172
7	0.165	12	0.169	17	0.171	>60	0.173

Tests for Linearity

$$U = \frac{\frac{\sum_{i=1}^{m} T_i}{m} - \frac{T}{2}}{T\sqrt{\frac{1}{12m}}}$$

At the 5% level of significance for $m \geq 4$, a positive score greater than 1.96 indicates that we can be more than 95% confident of an upward trend; for example, deterioration (and less than -1.96 indicates that we can be more than 95% confident of a downward trend; i.e., improvement). (Equivalent values for 90% confidence are +/- 1.645.) If the last event/failure coincides with the end of the observation period, use $m - 1$ instead of m in all three places in the formula. (Applying this formula to our sad face data set shows statistically significant (>95% confidence) of deterioration, with $U = 2.24$, agreeing with eyeball analysis). Rausand and Hoyland [16] generalized this framework by adding the homogeneity concept to analyze multiple repairable units. The Laplace trend test for multiple (combined) repairable units is:

$$U = \frac{S_1 + S_2 + \ldots + S_m - \frac{1}{2}\sum_{i=1}^{k} N_i X_i}{\sqrt{\frac{1}{12}\sum_{i=1}^{k} N_i X_i^2}}$$

where S_j is the sum of the failure times for the jth products; N_j is the number of failures for product j (noting to reduce by 1 for failure truncation); and X_j is the observation period for product j (or the last failure time for failure truncation). The no trend (null) hypothesis is rejected at significance level α if $|U| > Z_{\alpha/2}$.

▶ *The Mann Test.* Note that if the Laplace test (and the Military Handbook test) reject the no-trend hypothesis, the data set does not follow a HPP; yet the data set can still be trend-free (e.g., following a Renewal Process (RP) which these tests are not appropriate for). The Mann test can however distinguish whether a data set follows a RP or a NHPP [17]. Therefore, we might apply the Laplace (or Military Handbook) test to determine if we should reject the HPP hypothesis, and apply the Mann Test to determine if we should reject the RP hypothesis. (If no trend is indicated, dependence could still be at play, e.g., through a Branching Poisson process, but we often ignore this possibility and assume a RP or HPP is occurring. Similarly, if a trend exists, a trend renewal process could be at play, but we often ignore this possibility and assume a NHPP is occurring.)

- MIL-HDBK-189 Test (pp. 68–69).
- Anderson-Darling Test.

Cost Models of Repairable Items

The principal class of cost models for repairable items is concerned with deteriorating items, and consider only economic costs, not safety considerations.

Minimal Repair Policies Based on Repair Cost Limits

In many real world situations, a repair limit approach is used, using the principle that no more should be spent on the repair of an item than the item is worth. In this approach, the decision to repair or replace is based on the cost of repair compared to predetermined repair cost limits. If the cost of repair is less than this limit, the item is repaired; otherwise, the item is replaced or overhauled.

An Age-Dependent Minimal Repair Policy for Deteriorating Repairable Systems

An alternative maintenance policy assumes that after each failure the item is only minimally repaired and calls for planned replacement after some prespecified number of item operating hours, regardless of the number of intervening failures. This power law model, besides being mathematically tractable, is a good representation of many real repairable items. When the ROCOF is specifically of the form: $m(t) = abt^{b-1}$ (equivalent to this book's formula for number of failures $M(t) = at^b$) the optimal economic life replacement interval estimate (or overhaul if overhaul returns the item to same as new condition) is given by:

$$\left[\frac{scheduled\ system\ replacement\ cost}{a(b-1)(minimal\ repair\ cost)} \right]$$

where both costs are assumed constant. (See Barlow and Hunter [18].) Constraints and limitations of this and many other models include:

- Assumes instantaneous repair (consequently, includes only the minimal repair cost and the cost of replacing the system at failure, not other downtime costs).

- Assumes total repair costs are linearly related to the number of failures, and total equipment maintenance cost is a simple sum of the repair and replacement costs. (However, actual costs may interact with each other.)

- Assumes that the interarrival times of the new system can be modeled by the same point process as that of the replaced system. However, if times between system replacements are very long, then the replacement system may have different reliability characteristics.

Adaptive Cost Policies

Most cost models do not have a built-in procedure for revising the repair policy on the basis of information on the observed failure pattern of the system. Bassin [19] however adopted a Bayesian approach of this power law form in developing an adaptive maintenance policy where model parameters and the optimum replacement interval are continuously updated. This approach is quite cost effective because changes in the replacement interval can be immediately made on the basis of the updated estimate of net savings.

References

[1] Ascher, H., and H. Feingold, *Repairable Systems Reliability: Modeling, Inference, Misconceptions and Their Causes*, New York: Marcel Dekker, 1984

[2] Crow, L. H., "Estimation Procedures for the Duane Model," *ADA 019372*, 1972, pp. 32–44.

[3] Duane, J. T., "Learning Curve Approach to Reliability Monitoring," *IEEE Trans.*, A-2, 1964, pp. 563–566.

[4] Cox, D. R., "Regression Models and Life Tables (with Discussion)," *J. Roy. Stat. Soc.*, Ser B, Vol. 34, 1972, pp. 187–220.

[5] Kijima, M., and N. Sumita, "A Useful Generalization of Renewal Theory: Counting Process Governed by Non-Negative Markovian Increments," *Journal of Applied Probability*, Vol. 23, 1986, pp. 71–88.

[6] Guo, H., W. Zhao, and A. Mettas, "Practical Methods for Modelling Repairable Systems with Time Trends and Repair Effects," *IEEE*, 2006, pp. 182–188.

[7] Peng, Y., Y. Wang,Y. Zi, K. Tsui, and C. Zhang, "Dynamic Reliability Assessment and Prediction for Repairable System with Interval-Censored Data," *Reliability Engineering and System Safety*, Vol. 159, 2017, pp. 301–309.

[8] Blumenthal, S. B., J. A. Greenwood, and L. H. Herbach, "Superimposed Nonstationary Renewal Processes," *J. of Appl. Prob.*, Vol. 8, 1971, pp. 184–192.

[9] Reactor Safety Study, *An Assessment of Accident Risks in U.S. Commercial Nuclear Power Plants*, US Atomic Energy Commission, Washington D.C., 1974.

[10] Kujawski, G. J., and E. A. Rypka, "Effects of On-Off Cycling on Equipment Reliability," *Annual Reliability and Maintainability Symposium*, IEEE-77 CH1308-6R, 1978, pp. 225–230.

[11] Prentice, R. L., B. J. Williams, and A. V. Peterson, "On the Regression Analysis of Multivariate Failure Time Data," *Biometrika*, Vol. 68, 1981, pp. 373–379.

[12] You, M., and G. Meng, "Updated Proportional Hazards Model for Equipment Residual Life Prediction," *International Journal of Quality and Reliability Management*, Vol. 28, No. 7, 2011, pp. 781–795.

[13] Cox, D. R., "Regression Models and Life Tables (with Discussion)". *J. R. Statist. Soc.* B S4, 1972, pp. 187–220.

[14] IEC 61164-2008, *Reliability Growth – Statistical Test and Estimation Methods*.

[15] Yanez, M., F. Joglar, and M. Modarres, "Generalized Renewal Process for Analysis of Repairable Systems with Limited Failure Experience," *Reliability Engineering and System Safety*, Vol. 77, No. 2, 2002, pp. 167–180.

[16] Rausand, M., and A. Høyland, *System Reliability Theory: Models, Statistical Methods, and Applications*, Hoboken, NJ: Wiley, 2004.

[17] Garmabaki, A. H. S., A. Ahmadi, J. Block, H. Pham, and U. Kumar, "A Reliability Decision Framework for Multiple Repairable Units," *Reliability Engineering and System Safety*, Vol. 150, 2016, pp. 78–88.

[18] Barlow, R. E. and L. Hunter, "Optimum Preventive Maintenance Policies," *Operations Res.*, Vol. 8, 1960, pp. 90–100.

[19] Bassin, W. M, "A Bayesian Optimal Overhaul Interval Model for the Weibull Restoration Process," *J. Amer. Stat. Soc.*, Vol. 68, 1973, pp. 575–578.

CHAPTER 3

Contents

Introduction

Sample Statistics

Confidence Intervals and Nonparametric Analysis

Histograms

Cumulative Plots

Parametric Analysis Usually Follows Nonparametric Analysis

Appendix 3

Kaplan-Meier Method

Nonparametric Data Analysis

Introduction

This chapter and later chapter assumes our data set is IID.

Data analysis can take two general approaches: parametric and nonparametric. With parametric analysis, a statistic can be used to estimate (infer) the value of a parameter, or to validate a parametric model.

> NOTE: A fundamental question of statistics is, given a data sample, how do we infer the properties of the underlying distribution, and what confidence do we have in our estimates (inferences) of those properties?

An estimator is called unbiased if the mean of all possible values is equal to the parameter estimated. The sample mean is an unbiased estimator for the population (as a result of the Central Limit Theorem). However, an example of a biased estimator is the sample standard deviation (i.e., $\mu_s \neq \sigma$). An estimator for a parameter is called more efficient than another if it requires fewer samples to obtain an equally good approximation.

The other general approach to analyzing data is nonparametric analysis, which makes no assumption about the distribution from which the sample is drawn (we might say we let the data set speak for itself).

This chapter outlines common nonparametric analysis techniques:

- Sample statistics;
- Histograms;
- Cumulative plots based on rank statistics.

Chapter 6 extends this discussion on rank statistics to outline a particular technique that, while strictly parametric, has great utility as a quasi-nonparametric tool to provide insight into other appropriate parametric models if so desired—that technique is Weibull plotting.

Sample Statistics

Some sample statistics can be obtained without referring to a specific underlying probability distribution. The most common sample statistics measures central tendency characteristics. The word average as a measure of central tendency is seldom used formally by statisticians because it is ambiguous. The median (middle number when numbers are sorted by size) and mode (most frequently appearing number) are both measures of central tendency. However, mean or expected value is most often used.

Mean

For a set of x values of a sample numbering n, we can determine the mean (written as x-bar and also called expected value $E(x)$ or first moment) by simply adding all the values and dividing this sum by the number of values. This is written in mathematical shorthand as follows (noting Σ means summation):

$$\bar{x} = \sum_{i=1}^{n} \frac{x_i}{n} = \frac{1}{n} \sum_{i=1}^{n} x_i$$

If the values represented the entire population rather than a sample, the population mean symbol, u, would be used instead of $\bar{x}$ but the calculation is the same.

In Excel, use the AVERAGE function to calculate the mean.

MTTF and MTBF

Mean life or MTTF is the average time to failure of identical items operating under identical conditions. For a particular set of failure times, the mean life is obtained by calculating the mean of the times to failure. If this data set represents a random sample of the entire population of interest, this value serves as an estimate for the actual mean life (of the population) and is sometimes denoted with a hat (e.g., $\widehat{MTTF}$) to denote it is an estimate. MTBF strictly refers to repairable items that are promptly repaired after failure, but, in practice, $\widehat{MTBF}$ often refers to both repairable and nonrepairable items.[1]

In some situations, exact failure times are not known. Where n repairable items are tested for a given amount of time m (say, hrs) and promptly replaced, another formula for calculating mean life is:

$$\widehat{MTBF} = \text{(total time tested)} / \text{(number of failures, } r\text{)} = nm/r$$

However, readers should exercise care in the use of the terms MTTF and MTBF, since these terms are usually used when the failure rate is constant[2], or assumed so, which this formula necessarily assumes but which may not be the case in practice. Indeed, while many formulas are based on this simplifying assumption, not validating the assumption can lead to gross estimation errors.

MTTR

MTTR, defined as the average time taken to return an item to operational status, can also be calculated using the general summation formula for calculating a mean. However, many other downtime actions, besides pure repair, can be required to return an item to operational status. Therefore, in defining MTTR, often it is necessary to describe exactly what is included in the maintenance or repair action (e.g., time for: testing, diagnosis, part procurement, teardown, rebuild, verification, and administration).

Standard Deviation

The range (maximum value minus minimum value) and variance (or second moment) are both measures of dispersion or spread. However, standard

---
1. MTTF and MTBF are often denoted with the letter m or the Greek θ (theta).
2. The underlying failure distribution here is the exponential distribution.

deviation (the square root of the variance) is most often used (since it has the same units as the mean and is less sensitive to outliers, among other statistical advantages).

For a population, the standard deviation calculation formula is a little more complicated than that for the mean, but you should be able to see that the formula is based on how far away (dispersed) each data point is from the mean:

$$\sigma = \sqrt{\frac{\sum (x - u)^2}{N}}$$

where u is the population mean and N is the number of values in the population.

For the standard deviation of a sample, the formula is slightly different:

$$\hat{\sigma} = \sqrt{\frac{\sum (x - \tilde{x})^2}{n - 1}}$$

where $\bar{x}$ is the sample mean and n is the number of values in the sample. Note, the only practical difference in these formulas is that the divisor in the fraction is the number of values in a population, but one less than the number of values in a sample (this can be shown to provide a better estimate[3]).

In Excel, use the STDEV.P function to calculate the population standard deviation, and the STDEV.S function to calculate the sample standard deviation.

When using the standard deviation function on a calculator, take care to use the appropriate key for either population or sample standard deviation. Unfortunately, labeling among calculator manufacturers is not universal.[4]

Confidence Intervals and Nonparametric Analysis

A mean and standard deviation of a sample can be used to estimate the mean and standard deviation of a population. They are point estimates because they yield a single number, with no indication as to how much in error that number may be. More complex analysis provides confidence intervals to characterize the accuracy of a point estimate. Unfortunately though, with the exception of the mean (and only then for larger sample

---

3. It removes statistical bias, since $\bar{X}$ is estimated from the same sample.
4. If in doubt, remember that, for the same data, the sample standard deviation will always be larger than the population standard deviation (because of the division by n-1, rather than n).

sizes of, say, more than 30), confidence intervals can only be obtained after determining the form of the distribution. Therefore, confidence intervals apply to parametric analysis[5] and no distribution-free (nonparametric) confidence intervals exist for standard deviation, or other statistics, other than the mean.[6]

NOTE: No-failure confidence: While strictly derived from a parametric method (i.e., binomial analysis), the following formula is sometimes referred to as providing a nonparametric estimate of the number of test items needed to assure a particular reliability lower bound with a given confidence level, if subsequent testing of the n items produces no failures:

$$n = \frac{\ln(1 - Confidence\ Level)}{\ln R_{Lower}}$$

See binomial distribution in Chapter 10.

Histograms

Histograms are perhaps the most common form of nonparametric analysis. Histograms can be constructed as follows:

- Calculate the range of the data (i.e., maximum value minus minimum value).

- Divide the range into appropriate intervals. If too few intervals are used, lack of resolution obscures the nature of the distribution; if too many are used, large fluctuations in frequency can do the same. While no precise rule exists, Sturges' Rule can help (particularly for sample sizes less than, say, 200):

    number of intervals or 'bins' = $1 + 3.3 \log_{10}$ (number of data points).

- Check that no data point falls on an interval boundary.[7]

- Plot the number of data points in each interval.

---

5. The Fisher matrix method is utilized by most commercially available software packages for confidence bounds but is mathematically intense.
6. The mean is the exception because of an important statistical principle: the Central Limit Theorem (CLT) (see Chapter 8 appendix).
7. To guarantee data points not falling on an interval boundary, we might start the first interval below the lowest data value at a point calculated by: ½ the least significant figure in both the data values and the interval size.

This process can be automated to a degree using Excel by first installing Excel's Analysis ToolPak (or equivalent), then using the Histogram tool.

Distribution-EZ

Alternatively, we can use the Excel analysis tool provided, Distribution-EZ, to provide histograms of a data set.

Cumulative Plots

Even if insufficient data prevent reasonable histogram construction, we can still approximate the graph of how the random variable (e.g., number of failures) accumulates over time (termed the empirical cumulative distribution function or CDF), through rank statistics. We can simply plot the empirical CDF, defined for n (IID) times between successive failure, as

$$F_n(x) = \frac{\text{number of failure} \leq x}{n}$$

by listing the samplings of the random variable in ascending order (i.e., rank them). The CDF at each value is then approximated by

$$\hat{F}(x_i) = \frac{i}{N}$$

where $i = 1, 2, 3, ..., N$ and $F(0) = 0$.

While simply plotting the cumulative number of failures over time as a proportion of the total failures is the essence of cumulative plots, cumulative plotting can become a little more complex:

- *Median ranks.* If the number of data points is, say, less than 15, the above technique has some shortcomings. In particular, we find that $F(x)=1$ for x_N does not allow for any possible subsequent greater values which we might expect if we had, say, 100 or 1,000 values. Therefore, this method might be improved by arguing that if a very large sample was obtained, roughly equal numbers of events would occur in each of the intervals between our data points, and the number of samples larger than x_N would likely be about equal to the number within one interval. Therefore, a better estimate might be considered from a mean rank formula:

$$\hat{F}(x_i) = \frac{i}{N+1}$$

Other statistical arguments can be made for slightly different estimates using the median rank, one of the most common being:

$$\hat{F}(x_i) = \frac{i - 0.3}{N + 0.4}$$

For large samples sizes (values of N) however, all forms of $\hat{F}(x_i)$ yield nearly identical results.

- *Rank statistics for interval censored data.* Where exact failure times are unknown except within an interval (i.e., interval-censored data), the median rank method is not used to estimate the CDF but is replaced with:

$$\hat{F}(t) = \frac{Cumulative\ number\ of\ failures\ by\ time\ t}{Sample\ size\ used\ in\ test}$$

However, since we do not know when a failure occurred within any given interval, the midpoint (median) of each interval is often used as the associated time value for the $F(t)$ estimate.[8]

- *Nonparametric rank statistics for multicensored data.* If we are concerned with a particular failure mode when multiple failure modes occur, the different (unstudied) failure mode times can be handled as censored with respect to the mode under study (i.e., counted as a censored unit rather than a failure). Data of this sort requires special handling and analysis (e.g., using an adjustment of the median rank method, the Kaplan-Meier method [1], or other methods). The Kaplan-Meier method is outlined in the appendix. (Chapter 6 covers handling censored data using a parametric method—Weibull plotting.)

Example: Disk Drive Failures

Table 3.1 provides time to failures in hours of 16 computer hard disk drives [2].

---

8. This empirical distribution function cannot be used with operating times without failures (i.e., right censored data) but, for repairable system contexts, such a censored time can occur only for the last observed time between successive failure for any given system copy.

Table 3.1
Computer Hard Drive Failure Times
(in Operating Hours)

7	49	235	320	380	437	493	529
12	140	260	320	388	472	524	592

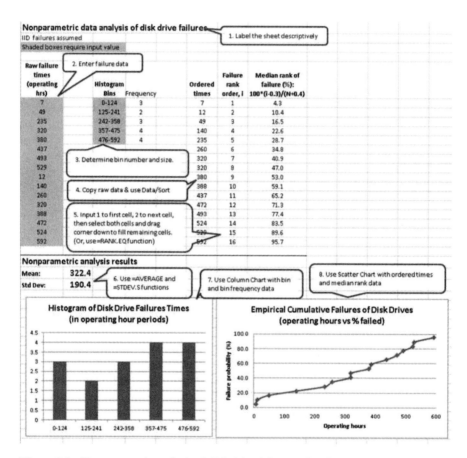

Figure 3.1 Nonparametric analysis of disk drive failures using Excel.

Figure 3.1 demonstrates creating a histogram and an empirical CDF of the data.

Nonparametric Tests

Since nonparametric methods do not require any strict distributional assumptions, they make fewer assumptions, they are more flexible, more robust, and applicable to nonquantitative data. However, in general, conclusions drawn from nonparametric methods are not as powerful as the parametric ones (but can be almost as powerful). Many nonparametric tests are based on empirical CDF statistics and analyze the ranks of a variable rather than the original values. Classes of nonparametric tests include: tests to make an inference about the median of a population from one sample,[9] tests to compare either the location parameters (e.g., medians) or the scale parameters of two independent population samples,[10] and nonparametric tests used in this book to determine whether two or more data sets have identical distributions, including the Chi-square goodness-of-fit test (see Chapter 4 appendix), the Kolmogorov-Smirnov test (see Chapter 4 appendix), and the Cramér-von Mises test (see Chapter 2 appendix).

Parametric Analysis Usually Follows Nonparametric Analysis

Parametric analysis usually provides better predictions and assessments based on the data than nonparametric analysis can provide. Parametric analysis includes both choosing the most appropriate probability distribution and evaluating the distribution parameters. However, initial nonparametric analysis is still recommended because it can provide valuable insight. Particularly when data paucity may not allow much basis for choosing a specific distribution, distribution-free or nonparametric techniques have much to recommend them. Even if a specific distribution is assumed, we can still benefit from using a distribution-free technique (e.g., to indicate how much the data imply and how much the model adds). Other factors guiding distribution choice include previous experience from similar circumstances, and the basis of the phenomena under investigation.[11]

Probability plotting both visually represents how well a particular distribution fits the data and provides point estimates of the distribution parameters. This technique is particularly valuable when:

▶ Very little data exists (which makes more classical methods of estimating parameters difficult);

---

9. Including the sign test and the Wilcoxon signed rank test. The Wilcoxon signed rank test requires that the distribution be symmetric (so the median is also the mean); the sign test does not require this assumption.
10. Including the median test and the Wilcoxon rank sum test (Mann-Whitney U test).
11. For example, the normal distribution may be suitable when the sum of many small effects is involved, the Weibull distribution may be suitable if it is a weakest link effect, and corresponding arguments can be made for other distributions.

- Not all events have occurred (e.g., not all products being tested have failed—termed right-censored data), which is common in reliability analysis but poses problems for nonparametric analysis.

Because of the flexibility of the Weibull distribution, Weibull plotting is frequently used initially. (See Chapter 6.)

APPENDIX 3

Kaplan-Meier Method

From lifetime data, the Kaplan-Meier reliability and hazard rate estimator equations are as follows: The Kaplan-Meier formulation is:

$$CDF\,(or\,Unreliability), \hat{F}(t_i) = 1 - \prod_{j=1}^{i} \frac{n_j - r_j}{n_j}$$

$$Reliability), \hat{R}(t_i) = 1 - \hat{F}(t_i) = \prod_{j=1}^{i} \frac{n_j - r_j}{n_j}$$

$$Hazard\,rate, \hat{h}(t_j) = \frac{r_j}{n_j \times \Delta t_j}$$

where $i = 1, \ldots$ (total number of data points); n is the total number of units; r_j is the number of failures in interval j; Δt_j is the time taken for r_j failures; and operating units in the interval j, $n_j = n - \sum r_j$. Chapter 4 outlines the reliability and hazard rate representations of failure data.

These equations might be better understood using a worked example (from Kalaiselvan and Bhaskara [3]), noting that the Kaplan-Meier method is particularly suited for censored data but, when applied to data without censoring, actually degrades to the rudimentary expression i/n for which the median rank method provides a better estimate.

Example: Nonparametric Analysis of Capacitor Reliability

A time to failure (in hours) data set obtained from accelerated testing of 50 capacitors is at Table 3.2:

Using the Kaplan-Meier method, values of capacitor hazard rate and reliability can be calculated as shown in Table 3.3:

Readers are encouraged to develop the full table (using Excel). Plotting hazard rate values from the full table shows hazard rate increasing over time. Similarly, plotting reliability values from the full table shows reliability reducing reasonably steadily from the first failure.

NOTE: This accelerated life test was conducted with increased temperature and voltage stresses applied as follows: 100 volts and 150 degrees Celsius (423K) compared to expected actual conditions of 50V and 75 degrees Celsius (348K)). Results were converted to actual expected time to failure under expected conditions using the Prokopowicz and Vaskas (P-V) empirical equation:

Table 3.2
Capacitor Failures in Hours from Accelerated Life Test

800	900	970	1030	1170	1270	1400	1510	1620	1690
810	910	990	1070	1200	1290	1420	1530	1640	1700
830	930	1000	1100	1220	1320	1430	1550	1650	1720
840	940	1010	1110	1230	1340	1450	1580	1660	1740
870	950	1020	1130	1250	1370	1470	1600	1670	1770

Table 3.3
Demonstration of the Kaplan-Meier Method

Data point (i)	Time to failure (t_j)	No. of failures (r_j)	No. units at beginning of observed time (n_j)	Hazard rate $\left(\frac{r_j}{n_i \times \Delta t_j}\right)$	$\left(\frac{n_j - r_j}{r_j}\right)$	Reliability $\prod\left(\frac{n_j - r_j}{r_j}\right)$
1	0 (start time)	0	50	0	1 (default if divide by zero)	1
2	800	1	50	0.000025	0.98	0.98
3	810	1	49	0.0000252	0.979592	0.96
...	...	...	...	...	...	...
50	1740	1	2	0.000287	0.5	0.02
51	1770	1	1	0.000565	0	0

$$\frac{t_1}{t_2} = \left(\frac{V_2}{V_1}\right)^n \exp\left[\frac{E_a}{k}\left(\frac{1}{T_{1abs}} - \frac{1}{T_{2abs}}\right)\right]$$

where values of $n = 2$, E_a for dialectric wearout = 0.5 eV, and k (Boltzmann's constant) of 8.62×10^{-5} eV were used (See Chapter 4 appendix for other accelerated life models). This gave an expected MTTF of over 11 years (illustrating why accelerated testing was preferred!).

Readers are also encouraged to consider how this table would change if, say, a single data point was censored (i.e., the capacitor ran for the stated number of hours without failing so the applicable line would have $r_j = 0$).

References

[1] Kaplan, E. L. and P. Meier, "Nonparametric Estimation from Incomplete Observations," *J. Amer. Stat. Soc.*, Vol. 53, 1958, pp. 457–481.

[2] Pasha, G. G., M. Shuaib Khan, and A. H. Pasha, "Empirical Analysis of the Weibull Distribution for Failure Data," *Journal of Statistics*, Vol. 13, No. 1, 2006, pp. 33–45.

[3] Kalaiselvan, C., and L. Bhaskara Rao, "Comparison of Reliability Techniques of Parametric and Non-Parametric Method," *Engineering Science and Technology,*, Vol. 19, 2016, pp. 691–699.

CHAPTER 4

Contents

CDF, PDF, and Hazard Rate

Bathtub Curve for Nonrepairable Items

Common Reliability Distributions

Failure Modes

Appendix 4

Reliability of Electronic Systems

Intermittent Failures

Stresses and Product Life Acceleration

Chi-Square (χ^2) Goodness-of-Fit Test

Kolmogorov-Smirnov (K-S) Goodness-of-Fit Test

Probability Distribution Representations

This chapter outlines the basic concepts and techniques involved in analyzing the times to failure of IID data, such as nonrepairable parts subject to the same usage and environmental conditions.

CDF, PDF, and Hazard Rate

Assume that we put a number (n) of (nominally) identical items on test under identical conditions, in such a way that no item is affected by any other item. In doing so, we assume that the p times to failure are independent samples from the same distribution function (i.e., IID). Note we replicate events to estimate the probability of a single event occurring—the failure of a single item (by time t, as a function of t).

We will use natural estimators to illustrate the three commonly used descriptions of the data: (cumulative) distribution function (CDF), (probability) density function (PDF), and hazard rate (see Figure 4.1):

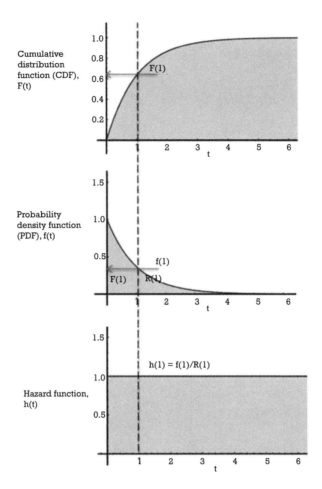

Figure 4.1 Different representations of the same phenomenon (example uses exponentially distributed failure times, and examines t=1).

CDF

In reliability engineering, a CDF provides the probability that a time to failure (of random variable T with a given probability distribution) will have a value less than or equal to a particular value (t). (Written mathematically as $F_T(t) \equiv Pr\{T \leq t\}$.) When each part tested operates until it fails, a natural estimator[1] for the CDF is

$$\frac{number\ of\ failure\ times \leq t}{n}$$

---

1. This is the empirical distribution function.

As a standard procedure in analyzing IID data, we reorder such data by magnitude (i.e., time to failure, to form order statistics), based on the definition of the CDF. (Reordering IID data by magnitude loses no information since no information is contained in the sequence of instants in time at which failures occur.) Therefore, imagine we (independently) test to failure, say, 1,000 nominally identical parts (or widgets) under identical conditions. As shown in the previous chapter, creating a scatterplot of times to failure vs. failure number, and then converting the failure number to a percentage of the total would show a natural estimate for the CDF of the widget reliability[2] (under the test conditions). If we measure a very large number, the function tends to a curve.[3] The value of a CDF necessarily eventually reaches 1.0 or 100%.

In reliability engineering we are concerned with the probability that an item will survive for a stated interval (0 to t, where t is the most suitable life unit, for example, time, cycles, distance, etc.). This is the reliability, given by the reliability function $R(t)$.[4] It follows that the reliability is simply 1—CDF.

PDF

Instead of a CDF representation of the widget failures, suppose we divide the x-axis (times to widget failure) into many intervals and count the percentage of failures in each interval. This histogram gives a natural estimator for the relative likelihood for a time to failure (T) to take on a given value (t). If we proceed to measure an increasingly large number and further reduce the measurement interval, the histogram tends to a curve which describes the population probability density function (PDF), $f(t)$.

NOTE: Mathematically, the PDF $f_x(x) \equiv F_x'(x)$ (i.e., the derivative of the CDF) and a natural estimator for the PDF is

$$\hat{f}_X \equiv \frac{1}{n} \frac{p(x) - p(x + \Delta x)}{\Delta x}$$

where $p(x)$ is the number of parts that survive to x and $p(0) = n$.

All values of a such a histogram, and the area under a PDF necessarily adds to 1.0 or 100%.

---

2. More correctly, unreliability
3. Mathematically, $F(t) = \int_{-\infty}^{t} f(t)dt$ (i.e., the integral of the PDF).
4. In the biomedical community, reliability is termed survivability, denoted as $S(t)$.

> **NOTE:** Joint and marginal distributions: When there is more than one random variable (which may or may not be independent), similar considerations apply (i.e., the PDF is $> = 0$ for x, y from minus infinity to infinity) and the volume under the PDF (i.e., the double integral of $f(x, y)$)= 1. However, the probability that 2 or more random variables fall within a specified subset of the range is the double integral over the respective ranges; the marginal PDF, $g(x)$ of (say) y is given as the integral of the joint function wrt x; and the conditional probability $Pr(x|y) = f(x,y) / g(y)$, $g(y) > 0$. If x, y are independent, then $f(x|y) = f(x)$ and $f(x,y) = g(x) h(y)$.

Hazard Rate

A third representation of widget failures considers the relative likelihood for a time to failure (T) to take on a given value (t) as a proportion of the number of widgets remaining at that time. We call this the hazard rate $h(t)$, the number of failures per unit time, per number at risk at t.

> **NOTE:** Mathematically, the hazard rate
>
> $$h_X(x) \equiv \frac{f_X(x)}{1 - F_X(x)} = \frac{f(x)}{R(x)}$$
>
> A natural estimator for the hazard rate is:
>
> $$\widehat{h}_X(x) \equiv \frac{1}{p(x)} \frac{p(x) - p(x + \Delta x)}{\Delta x}$$

The hazard rate is also known as hazard function, mortality rate, or instantaneous failure rate.

The normalizing factor denominator—the number of items surviving at that time—represents the only difference between the hazard rate and PDF estimators. However, this is a very important difference. Using people as an example, in any particular time period, very few centenarians die compared to the number of people aged 50 who die (in absolute terms). Nevertheless, since the number of people surviving to 100 is very much less than the number of people surviving to 50, the relative rate, deaths per unit time, per number of people at risk at age 100 (i.e., deaths per person years of exposure to the risk of death) is very high and increasing rapidly compared to the relative rate of deaths of people aged 50.

Bathtub Curve for Nonrepairable Items

Since a probability distribution can be represented in many equivalent ways (see Figure 4.1), any representation can be converted to any other representation.

NOTE: A conversion table is:

	$f(t)$	$F(t)$	$R(t)$	$h(t)$
$f(t)=$		$F'(t)$	$h(t)R(t) - R'(t)$	$h(t)R(t)$
$F(t)=$	$\int_{-\infty}^{t} f(x)dx$		$1 - R(t)$	$1 - e^{-\int_0^t h(y)dy}$
$R(t)=$	$f(t)/h(t)$	$1 - F(t)$		$e^{-\int_0^t h(y)dy}$
$h(t)=$	$f(t)/R(t)$	$f(t)/[1 - F(t)]$	$\dfrac{d(-\ln R(t))}{dt}$	

The representation that should be used is the one that is most useful for the context.

Once reliability engineers assess natural estimator failure data, they typically then try to fit this data set into standard reliability engineering statistical models to make better predictions and assessments based on the data (usually for different bathtub curve phases, as outlined in the next section). They do so by assessing goodness-of-fit of potential models, estimating model parameters, and developing confidence intervals for the model parameters.

Bathtub Curve for Nonrepairable Items

The behavior of hazard rates with time can be quite revealing, and the hazard function gives rise to the widely-used concept of the bathtub curve, which aligns with three broad causes of reliability performance variability. Because the bathtub curve concept characterizes living creatures as well as engineering items, much of the terminology comes from those studying human mortality. See Figure 4.2.

The three phases of the bathtub curve are as follows:

▶ Variability in manufacturing and maintenance (infant mortality or burn-in). Generally, deficiencies in manufacturing or transportation lead to failures concentrated early in an item's life, giving rise to the decreasing hazard rate portion of the bathtub curve referred to as infant mortality. Infant mortality hazard rates can be high but rapidly decreasing as producers identify and discard defective parts and fix

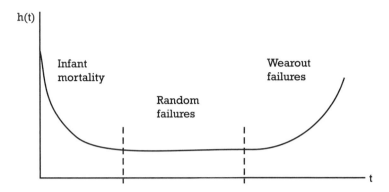

Figure 4.2 Classical nonrepairable item bathtub curve.

early sources of potential failure such as handling and installation errors.

NOTE: Sometimes, deliberate stress or overstress can be applied to cause weak items to fail (high stress burn-in of electronic devices, proof-testing of pressure vessels, etc.), leaving a population with a left-truncated strength distribution. Such tests not only destroy weak items, but may also weaken good items. Therefore, it is not economical to improve population reliability by this screening method in situations of a low safety margin and a wide load distribution. The options left include increasing the safety margin by increasing the mean strength (which might be expensive) or curtailing the load distribution (e.g., through devices such as current limiters and fuses in electronic circuits or pressure relief valves and dampers in pneumatic and hydraulic systems).

- Variability in the operating environment (useful life). Extremes in the operating environment(such as power surges, mechanical impact, temperature fluctuations, moisture variation, vibration, nails on roads etc.) generally occur randomly throughout an item's life, leading to a hazard rate independent of the item's age (rather than any inherent defect in the item). This period is commonly referred to as the useful life, accounting for the generally flat or constant hazard rate portion of the bathtub curve. Useful life failures can be reduced by making designs more robust or operating the product in a reduced load environment.

- Wear out (aging). Failures from wear out (broadly defined to include cumulative effects from wear, corrosion, embrittlement, fatigue cracking, diffusion of materials, etc.) most frequently occur toward the end of an item's life, giving rise to the increasing hazard rate portion of the bathtub curve. Decisions to replace items is often based on the start of rapidly increasing hazard rates [1] (specifying the item's design life).

While useful as a concept, note that the reliability of many items may not follow a bathtub curve [2].

NOTE: Muth [1] showed that a diminishing mean residual life (MRL) function—the expected additional time to failure given survival to x—was a more meaningful indicator of wear out than the hazard rate of a distribution. The maintenance that should be conducted (both amount and type) depends on the cost of failure in its broadest sense, particularly safety implications and the relative cost of downtime, both direct (e.g., maintenance) and indirect (broader support costs and lost operations). Further, while these theoretical considerations assume that maintenance does not introduce any defects, because human reliability is less than perfect, maintenance can tend to introduce an infant mortality region back into the bathtub curve at each maintenance event (i.e., decreasing overall reliability). Maintenance-induced failures may be a primary cause of common-cause failures in redundant safety critical systems (e.g., with multiple aircraft engines, if the wrong lubricant is put in one engine, it is likely to be put into a second also).

NOTE: Nolan and Heap [2] claimed to show hazard rate patterns of commercial aircraft components tended to fall into one of six major groups, with 89% of the items analyzed showing no wear out zone. The two groups of components (6% of items) which together show clear wear out characteristics (which they called groups A and B) are associated with a great many single-celled or simple items (e.g., tires, reciprocating-engine cylinders, brake pads, turbine-engine compressor blades, and all parts of the airplane structure). Note however that since the horizontal axis in their publication represents age since manufacture, overhaul, or repair (e.g., of subsystems such as engines), this publication may mix hazard rate with ROCOF. (See Chapter 2.)

General characteristics of specific items in relation to the bathtub curve are:

- Computer and electronic products often have a long constant hazard rate useful life period, so the primary concern is with controlling the environment and external loading to reduce random failures, or to design items that can handle more environmental extremes (or both). (See the appendix.)
  - Products that are primarily mechanical in nature—such as valves, pumps, and engines—tend to have more of a monotonically increasing hazard rate after wear-in (infant mortality) (since the primary failure

mechanisms are wear, fatigue, corrosion, etc.). Therefore, the primary concerns are with estimating safe or economical operating lives and developing practical preventive maintenance schedules.

NOTE: With good manufacturing quality control (and/or burn-in), and careful preventive maintenance (including timely component replacement), repairable mechanical equipment can sometimes also be modeled adequately by the useful life portion of the bathtub curve. Conversely, if components in a complex repairable system are replaced as they fail, over time (e.g., when every part of a system has been replaced several times), the overall failure rate of the system will appear roughly constant, since the component failures will be randomly distributed in time as will the ages of the replacement components. The danger in this situation is to assess that, since the system hazard rate is constant, preventive maintenance will have no effect and is not warranted. In fact, in this situation, preventive maintenance applied to wear out components may reduce the overall system hazard rate dramatically. (See Chapter 2.)

- With software, the issue is with logical errors or oversights committed during design or establishing requirements. Even with extensive testing, it may be impractical (or impossible) to find and eliminate all errors before production. Thus, there may exist untested sets of inputs that will cause a system to malfunction, and these may occur randomly in time, equating to the useful life portion of the bathtub curve. However, with the common practice of finding and correcting bugs (i.e., redesign) both before and after software is released for use, this reliability growth can be modeled by referring to the infant mortality component of the bathtub curve (especially if occurring frequently).

- Increasing (failure inducing) stress levels increase the hazard rates of products at all phases of the bathtub curve. (See the appendix.)

Common Reliability Distributions

What (if any) distribution is appropriate to represent the data? Factors guiding distribution choice include:

- Whether the metrics of concern are countable (discrete) or measurable (continuous);
- Insight from nonparametric analysis;
- Previous experience from similar circumstances;

Common Reliability Distributions

- The basis of the phenomena under investigation;
- Whether a constant rate can be (reasonably) assumed;
- Goodness-of-fit for the data to particular distributions (see the appendix).

The most common distributions used in reliability analysis can be grouped as follows:

- Those where the random variable is measurable (i.e., continuous):
  - Weibull distribution;
  - Exponential distribution;
  - Normal distribution;
  - Lognormal distribution.
- Those where the random variable is countable (i.e., discrete):
  - Binomial distribution;
  - Poisson distribution.

The next several chapters outline these distributions for reliability analysis.

Distribution-EZ

If our random variable is measurable (continuous), we can use the Excel analysis tool provided, Distribution-EZ, to assess how well our data set is represented by the four continuous distributions listed. This tool plots a histogram of the data, provides goodness-of-fit assessments for each of the four continuous distributions listed, and calculates associated distribution parameters.

Sampling Distributions

Reliability analysis also utilizes other distributions that are not used directly as reliability or maintainability distributions, but as sampling distributions that relate to statistical testing, goodness-of-fit tests, and evaluating confidence. The most common (used in this book) are listed below but are not further individually explained:

- The Chi-squared (χ^2) distribution is very useful for goodness-of-fit tests and evaluating confidence;

- The (Student's) t distribution is used to obtain confidence intervals for the normal distribution mean (and other percentiles) based on sample data;

- The F distribution can be used to compare means of exponential distributions, and to compare variances of normal distributions;

- The beta distribution can be used to model the random behavior of percentages and proportions.[5]

NOTE: Other distributions—readers may come across the following less-used probability distributions in their reliability studies or research:

Gamma distribution. In reliability terms, the gamma distribution describes situations when a given number of partial failures must occur before an item fails. In statistical terms, the gamma distribution represents the sum of exponentially distributed random variables (so the exponential distribution is a special case of the gamma distribution). The gamma distribution can also be used to describe a decreasing hazard rate.

Extreme value distributions. In reliability work we are often less concerned with the distribution describing the bulk of the population than with the implications of the more extremes in the tails of the distribution. We often should not assume, because a measured value appears to be, say, normally distributed, that this distribution is necessarily a good model for the extremes when few measurements are likely to have been made at these extremes. Extreme value statistics are capable of describing these situations asymptotically (for maximum and minimum values) and are the limiting models for the right and left tails of the normal, lognormal, and exponential distributions (indeed, any distribution whose cumulative probability approaches unity at a rate that is equal to or greater than that for the exponential distribution). Notably, the relationship between the Weibull and lowest extreme value distribution is analogous to the relationship between the lognormal and normal distribution so, for lifetime distribution modelling (where failure times are bounded below by zero), the Weibull distribution is a good choice as the limiting distribution. For a system consisting of many components in series, where the system hazard rate is decreasing from $t = 0$ (i.e., bounded), a Weibull distribution will be a good model for the system time to failure, independent of the choice of component model.

Hypergeometric (discrete) distribution. The hypergeometric distribution models the probability of k successes in n Bernoulli trials from a population N containing m success without replacement. This distribution is useful for quality assurance

---

5. In Bayesian inference, the beta distribution is also the conjugate prior probability distribution for the Bernoulli, binomial, negative binomial and geometric distributions. See Chapter 13.

Failure Modes

and acceptance tests (e.g., for determining the probability of finding k objects in sample, given finite population N, with m items of interest, and sample size n). However, often the simpler binomial distribution (with $p = m/N$) is used in its place, since the function approaches a binomial function when the sample size is small relative to the population (n/N).

Degrees of Freedom

One input into (Excel) formula relating to these sampling distributions is degrees of freedom. Formally, degrees of freedom relate to the number of values in the final calculation that are free to vary. Throughout this book, the number of degrees of freedom to be used in (Excel) formula is often one less than the number of failures. (We have seen this in practice earlier with the calculation of sample standard deviation.)

Failure Modes

Reliability distributions can be used to model nonrepairable systems with one dominant failure mechanism. In many (most) real systems, however, failures occur through several different mechanisms, causing a distribution shape too complex to be described by any single one of the distributions described. For example, several failure mechanisms within a single structure such as a car tire can cause failure: defective sidewalls, puncture, and tread wear. Physically distinct components, such as the processor unit, memory, or disk drives of a computer, may fail in different dominant ways. In either case, provided the failures are independent, if we can separate the failures according to the mechanisms or components that caused them, we can treat the system reliability in terms of mechanisms or component failures, collectively known as independent failure modes. If the failure modes are independent,[6] we can calculate the system reliability as simply the product of independent failure mode reliabilities,

$$R_{sys}(t) = \prod_i R_i(t),$$

or system hazard rate as simply the sum of independent mode hazard rates, $h_{sys}(t) = \sum_i h_i(t)$. When independent failure modes can be approximated by constant hazard rates, the system hazard rate calculation can be further simplified by removing the time (life unit) dependency:

---

6. If failure modes or the failures of different components arise from a single cause, contributions can strongly interact and treating them as independent through using these formulas is not valid.

$$\lambda_{sys} = \sum_i \lambda_i \quad \text{or} \quad \frac{1}{\text{MTTF}} = \sum_i \frac{1}{\text{MTTF}_i}$$

NOTE: The assumption of constant hazard rates is the basis of parts count methods, such as described in MIL-HDBK-217F [3] and similar commercial standards. While MIL-HDBK-217F is rather dated, it has spawned a progeny of fundamentally similar handbook-based methods used to estimate reliability parameters including Telcordia SR-332, PRISM, RDF-2000, 217Plus, FIDES, and Siemens SN29500. The handbook methods have also been computerized and commercialized by numerous companies as software packages including Item Software, Reliasoft Lambda Predict, T-Cubed, ALD Reliability Software, Quanterion (217Plus), and Windchill Prediction.

Employing MIL-HDBK-217 methodologies and its progeny to predict reliability might seem easy and cheap, but all these handbooks and tools have been criticized as flawed and leading to inaccurate and misleading results. Discrepancies of several orders of magnitude have been observed across a wide range of devices and technologies. Inaccuracies are inherent to handbook-based reliability estimation methodologies because such estimations fail to adequately capture the cause-and-effect relationships, such as temperature factors, environmental and operating loading conditions, and evolving technologies. Better alternatives include using field data, test data, and data from similar technologies used in similar environmental and use conditions. (See Pandian et al. [4])

APPENDIX 4

Reliability of Electronic Systems

Integrated Circuits and Nonmicroelectronic Component Reliability

The manufacturing quality of modern complex integrated circuits is so high that defective proportions are typically less than ten per million (and there are no practical limits to the reliability that can be achieved with reasonable care). Discrete microelectronic components are generally also very reliable, though some do degrade intrinsically, including light-emitting diodes (LEDs), relays (mainly due to arcing progressively damaging the contact surfaces), and electrolytic capacitors. (If no voltage stress is applied, these capacitors degrade over time and then fail short-circuit when used; they must be reformed at intervals to prevent this. They are also damaged by reverse or alternating voltage.)

Competing against this force of increasing electronic systems reliability through manufacturing quality is a trend for new or more pronounced failure modes as semiconductor devices become ever smaller. Indeed, with commercial electronics moving into the sub-23-nm range, semiconductor sizes are having more distinct failure mechanisms, and the mean life of microcircuits is expected to reduce to below 10 years. (See Pandian et al. [4])

Key Stressors

Apart from selecting suitable components, circuit and system designers generally have little control over the design (and hence reliability) of electronic components. However, they can significantly influence system-level reliability through the following electronics key reliability practices:

- Transient voltage protection including from electrostatic discharge (ESD) and unregulated power supplies. Arcing is another voltage stress that can occur when contacts are opened, for example between brushes and commutators of motors and generators (arcing can be reduced by using capacitors across relay or switch contacts).

- Thermal design, noting that the widely held the cooler the better rule-of-thumb is erroneous. While controlling temperatures, particularly localized hot spots, may still be necessary, repeated large temperature changes can be more damaging than continuous operation at high temperatures.

- Electromagnetic interference/compatibility (EMI/EMC) (particularly important in high frequency digital systems). The main design techniques include filter circuits or decoupling capacitors, shielding circuits and conductors (by enclosing them in grounded, conductive boxes); balancing circuit impedances, and ensuring ground connections have low impedance paths back to the current source.

Cables, Connectors, and Solder Joints

In many modern electronic systems, many failures in service can be traced back to connectors, of many kinds but mainly at terminations or at stress/bend points. Interconnects such as solder joints and wire bonds can also be reliability weak links (e.g., from solder joint fatigue and cracking) especially in harsh environments.

Intermittent Failures

Sometimes more than 50% (up to 80%) of reported electronic system failures are diagnosed as no fault found (NFF) or retest OK (RTOK). This high rate may in part be due to the effects of intermittent failures, where the system performs incorrectly only under certain conditions, but not others. When the causes of the NFF/RTOK failures are not detected, the faulty units are not repaired, and can cause the same failure when reinstalled. This loop can generate high downtime, and high costs in (non)repair work and spares.

Intermittent failures are often caused by:

- Connectors that fail to connect (e.g., when under vibration or at certain temperatures).
- Broken circuit card tracks that are intermittently open circuit.
- Built-in test (BIT) systems that falsely indicate failures that have not actually occurred. (Alternatively, the diagnosis of which item has failed might be ambiguous, so more than one is replaced even though only one has failed. In these cases, multiple units are sent for repair, resulting in a proportion being classified NFF/RTOK. Returning multiple units can sometimes be justified, e.g., to spend the minimum time diagnosing the cause of a problem in order to return the system to operation as soon as possible.)
- Inconsistent test criteria (e.g., between the in-service test and the repair depot test).
- Human error or inexperience.
- Tolerance build-up effects between component parameters.

NFF/RTOK rates can be reduced by effectively managing the design relating to in-service test, diagnosis and repair. Stress screening can also reduce the proportion of failed items returned.

Stresses and Product Life Acceleration

Generally, higher stress levels shorten the expected life of products and increase their hazard rates at all phases of the bathtub curve. (See Figure 4.3.)

While no universal model exists that applies to all items, most accelerated life models are not linear models. The obvious exception is the simplest, when an item is used infrequently—simply speed up the rate of use. This technique might apply to toasters, washing machines, doors, some engines and motors, and switches. For example, automobile life testing may be compressed by leasing them as taxis, and home kitchen appliances can be tested in commercial restaurants (while some difference need to be accounted for, the data may be adequate for particular needs). However, the most common accelerated life models are as follows.

The Arrhenius model is often used to model thermal stress effects (e.g., of electronic components). Here, a life acceleration factor (A) is determined by:

$$A = e^{\frac{E_a}{k}\left[\left(\frac{1}{T_{low}}\right) - \left(\frac{1}{T_{high}}\right)\right]}$$

where temperature T is in units of Kelvin, k is Boltzmann's constant, and E_a is the activation energy specific to the failure mechanism, derived empirically (through testing). However, an approximation that can sometimes be used in the absence of knowing the activation energy is that every 10°

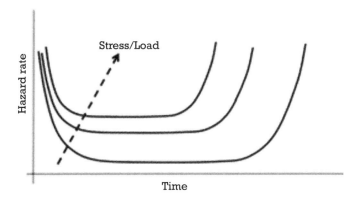

Figure 4.3 Effect of accelerated test on the bathtub curve.

centigrade increase in temperature (at the failure mechanism site) doubles the failure rate.

Example: Data Center Cooling

Given some estimates put total data center electricity consumption at 1.5% of global electricity production, and that half of this power is used for cooling, Li, Reger, and Miller studied the effect of cooling air temperature on integrated circuit memory module reliability, specifically the effect of inlet air temperature on (NetApp) storage server failure rates [5]. Conventional wisdom is that the cooler the ambient temperature, the better the reliability of electronic hardware. The study found that storage server failure rate did increase with inlet air temperature and that the data fitted well with the Arrhenius equation (with an activation energy of 0.22eV).

The more general exponential model, $Life = (constant)e^{-stressor}$ describes the rate at which a wide variety of chemical reactions take place, including metal corrosion, breakdown of lubricants, or diffusion of semiconductor materials, and may also be used with vibration or voltage as stressors.

The inverse power law model is also used to model many other accelerating forces. Here, a life acceleration factor is determined by:

$$A = \left\{ \frac{V_{high}}{V_{low}} \right\}^p$$

where V is the accelerating stressor (such as voltage) and p is a constant specific to the failure mechanism, derived empirically (through testing). Failure mechanisms of electronic components typically follow this relationship for dependence on voltage, electric motors follow a similar form for dependence on stresses other than temperature, and mechanical creep of some materials may follow this form.

These models are sometimes combined, as in the (simple) Eyring model with a life acceleration factor

$$A = \frac{Stress_{high}}{Stress_{low}} e^{constant\left[\left(\frac{1}{Stress_{low}}\right) - \left(\frac{1}{Stress_{high}}\right)\right]}$$

where the constant is derived empirically. This is used to model processes based on chemical reactions, diffusion, corrosion, or migration.

Chi-Square (χ^2) Goodness-of-Fit Test

Goodness-of-fit tests help determine whether a particular sample has been drawn from a known population. In their standard form they provide a level of confidence about whether a specific distribution with known parameters fits a given set of data. (Note that they assume not only that a distribution has been chosen, but that the parameters are known. In contrast, in probability plotting, we attempt to both estimate distribution parameters and establish how well the data fit the resulting distribution.)

The χ^2 goodness-of-fit test (also known as the χ^2 test for the significance of differences) is a commonly used and versatile test since it can be applied to any assumed distribution (provided that a reasonably large number of data points is available —desirably at least three data classes, or cells, with at least five data points in each cell). High values of χ^2 cast doubt on the hypothesis that the sample comes from the assumed distribution. The justification for this test is that, if a sample is divided into several cells, then the values within each cell would be normally distributed about the expected value if the assumed distribution is correct.

The Chi-square goodness-of-fit test statistic is:

$$\chi^2 = \sum_{i=1}^{k} \frac{\left(\text{Observed}_i - \text{Expected}_i\right)^2}{\text{Expected}_i}$$

This value is compared with a critical value of the Chi-square distribution, for a given desired confidence.

To get a feel for this test, let us construct Table 4.1 displaying the expected distribution if the sample exactly followed the historical or expected distribution.

Calculate the test statistic by summing the last column of the table. With $(n - 1)$ degrees of freedom (in this case there are 3 degrees of freedom) we use the Excel calculation of the appropriate χ^2 distribution,

Table 4.1
Chi-Square Goodness-of-Fit Test Statistic

Defective type	Observed frequency (O)	Expected frequency (E)	(O-E)²/E
Paint run	27	33.28	1.19
Paint blister	65	58.24	0.78
Decal crooked	95	87.36	0.67
Door cracked	21	29.12	2.26
Total	208		4.9

= CHISQ.DIST.RT(4.9, 3) to give 0.179 (or we could use the Excel function CHISQ.TEST to give this same result). (This is always a right-tailed test.) We want 90% confidence, so the significance level and critical value is 0.1 (i.e., 1-confidence level). Since 0.179 is larger than the significance level (critical value) of 0.1, we reject the (null) hypothesis that the sample comes from the expected distribution.

Note that any conclusion to not reject (based on goodness-of-fit tests) should not be inferred as accepting the (null) hypotheses. Such an inference is too strong when such tests only give evidence that the null hypothesis is false. (There may be better hypotheses, better fitting distributions.)

Kolmogorov-Smirnov (K-S) Goodness-of-Fit Test

The K-S test, sometimes paired with the χ^2 test, is also used extensively in reliability engineering to assess goodness of fit of a data set with a selected distribution (i.e., to decide whether a sample comes from a population with a specific distribution). When compared with the χ^2 test, the K-S test can give better results with relatively few data points. However, K-S test drawbacks include that it is generally considered to require complete samples that limit its use in many reliability analyses and it requires ungrouped data (i.e., specific times at which individual item failures occur). (Field failure data sets are more likely to be grouped.)

The procedure is: first, tabulate the ranked failure data (i.e., establish an empirical cumulative distribution function). For each data point (Y_i), calculate the positive value of the difference between cumulative rank value and the expected cumulative rank value for the assumed distribution ($F(x)$) using, as appropriate:

$$F(Y_i) - \frac{i-1}{N}, \text{ or } \frac{i}{N} - F(Y_i)$$

The K-S test statistic is the highest single value of these differences. Compare this value with the appropriate critical K-S value obtained from a table, for a given significance level (1-confidence level). (When the parameters of the assumed CDF are being estimated from the same sample, the critical K-S values are too large and give lower significance levels than are appropriate in the circumstances. To correct for this, the critical values should be multiplied by 0.7.) We do not reject the (null) hypothesis—that the sample comes from a population with the selected distribution—if the K-S test statistic is less than the K-S critical value.

Kolmogorov-Smirnov (K-S) Goodness-of-Fit Test

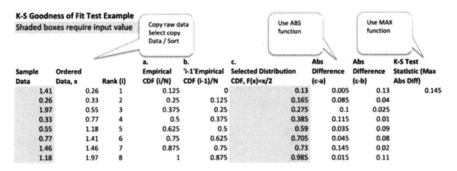

Figure 4.4 Developing a Kolmogorov-Smirnov test statistic using Excel.

Table 4.2
K-S Critical Values at 10% Significance

n=	1	2	3	4	5	6	7	8	9	10	11	12
	0.950	0.776	0.636	0.565	0.510	0.468	0.436	0.410	0.387	0.369	0.352	0.338
n=	13	14	15	16	17	18	19	20	21	22	23	24
	0.325	0.314	0.304	0.295	0.286	0.279	0.271	0.265	0.259	0.253	0.247	0.242
n=	25	26	27	28	29	30	31	32	33	34	35	>35
	0.238	0.233	0.229	0.225	0.221	0.218	0.214	0.211	0.208	0.205	0.202	$\frac{1.224}{\sqrt{n}}$

To get a feel for this test, let us construct an Excel spreadsheet to test the goodness-of-fit of some sample data to a uniform (PDF) distribution (from $x = 0$ to $x = 2$). See Figure 4.4.

Ninety percent confidence (10% significance) K-S critical values are at Table 4.2.

Since our K-S test statistic (0.145) is less than our critical value (0.410), we cannot reject the (null) hypothesis that the data set comes from our selected (uniform) distribution.

References

[1] Muth, E. J., "Reliability Models with Positive Memory from the Mean Residual Life Function," *Theory and Application of Reliability* (C. P. Tsokos and I. M. Shimi, eds), New York: Academic Press, Vol. 2, 1977, pp. 401–435.

[2] Nolan, F. S., H. F. and Heap, *Reliability-Centered Maintenance, United Airlines for US DoD*, Report AD-A066-579, 1978.

[3] MIL-HDBK-217F, Military Handbook, *Reliability Prediction of Electronic Equipment*, Washington DC, Department of Defense, Dec. 1991.

[4] Pandian G. P., D. Das, C. Li, E. Zio, and M. Pecht, "A Critique of Reliability Prediction Techniques for Avionics Applications," *Chinese Journal of Aeronautics*, Vol. 31, No. 1, 2018, pp. 10–20.

[5] Li, J., B. Reger, and S. Miller, "Leveraging Big Data to Improve Reliability and Maintainability," *IEEE*, 2015.

CHAPTER 5

Contents

Introduction

Distribution-EZ

Applying the Weibull Distribution Model

Weibull (Continuous) Distribution

Introduction

The Weibull distribution is used often by reliability engineers because it can model all three stages of the bathtub curve (i.e., decreasing, constant, and increasing hazard rates)—though not at the same time—as well as approximate other life distributions. (See Figure 5.1.) The Weibull distribution is particularly justified for situations where the weakest link is responsible for failure. However, even in situations where the underlying distribution is not explicitly known but the failure mechanism arises from many competing flaws, the Weibull distribution often provides a good empirical fit to the data. It is especially applicable to strength of materials.

> **NOTE:** Advantages of the Weibull distribution include:
>
> - The Weibull distribution is the log of the smallest extreme value distribution. Accordingly, the Weibull distribution is sometimes referred to as a competing risk model, with many potential faults all competing to be the first to cause failure (associated with the phrase "a chain is only as strong as its weakest link").

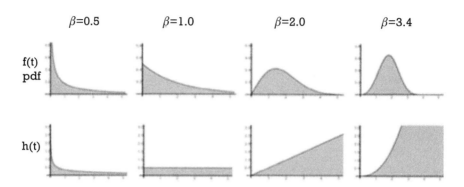

Figure 5.1 Flexibility of the Weibull distribution: With different shape parameters (β), the hazard rate h(t) can replicate any portion of the bathtub curve, and the PDF can approximate the exponential, lognormal, and normal distributions. (Weibull scale parameter is 2 in all figures.)

- Its domain lies only on the positive real line (like the lognormal distribution but unlike the normal distribution).
- The Weibull distribution exhibits the closure property (like the exponential distribution). For a series of units when the units have the same shape parameter β, the series reliability can be determined:

$$R_{series} = e^{-t^{\beta}\left\{\sum_{i=1}^{n}\frac{1}{(\eta_i)^{\beta}}\right\}}$$

The Weibull distribution has two parameters: a shape parameter (beta) $\beta \geq 0$, and a scale parameter[1] eta, $\eta \geq 0$ (though the symbols for these parameters differ among different texts).

▶ The shape parameter β provides for the flexibility of the Weibull distribution: values between 0 and 1 provide a decreasing hazard rate reliability function, useful for describing the infant mortality portion of the bathtub curve. The Weibull PDF is also monotonically decreasing.

▶ A special case of the Weibull distribution is when $\beta = 1$, when the Weibull distribution formula simplifies into the single-parameter exponential distribution (with η = mean life), providing a constant hazard rate reliability function useful for describing the useful life portion of the bathtub curve.[2]

---

1. Also called characteristic life, the life at which 63.2 % of the population will have failed.
2. When β is very close to 1, the abrupt shift in the PDF value at t=0 can complicate maximum likelihood estimates (when β=0.999, f(0) = ∞, but when β=1.001, f(0) = 0).

Introduction

- When $\beta > 1$, we get an increasing hazard rate reliability function useful for describing the wearout portion of the bathtub curve.

- When β is around 2, the Weibull distribution approximates to the lognormal distribution (and has a linearly increasing hazard rate).

NOTE: The Weibull hazard rate linearly increases when $\beta=2$. This distributional form is known as the Rayleigh distribution, and is used in diverse applications including wind speed, light scattering, surface waves, and radial error of measurement. For β values between 1 and 2, the hazard rate increases but at a diminishing rate. For β values larger than 2, the hazard rate increases at an increasing rate. For example, β-slopes are usually around: 2 for ball bearing failures, 2.5 for rubber V-belts, and 2 – 4 for low cycle solder fatigue.

When determining whether a Weibull or a lognormal distribution should be used for right-skewed data, look at the physical failure mechanism. While the Weibull distribution is a competing risk or weakest link model, and applicable to strength of materials, the lognormal distribution is known as a slow fatigue or wear out model. A general rule is that for considering early failures (e.g., the time associated with the first 1% of failures) the Weibull distribution is more cautious (e.g., earlier times) and the lognormal distribution is more optimistic. (The Weibull distribution is associated with the smallest extreme value distribution.) Finally, since it is linked to the normal distribution, one possible way of checking whether a lognormal distribution should be favored over a Weibull distribution is to take the log of the data and observe whether the histogram resembles a normal distribution.

- When β is between 3 and 4, the distribution approximates to the normal distribution.[3]

- For β values beyond 3.5, the PDF becomes increasingly skewed to the left. Because the very high onset of failures and effective failure-free period associated with β values beyond 6 are not normally experienced in most applications, reliability engineers may become suspicious of such results without some plausible explanation.

NOTE: In engineering practice, failure mechanisms with a true failure-free period are not usual (except corrosion and fatigue). If the failures are caused by some extreme conditions (e.g., high electrical load or low bond strength), then extreme value distributions may be the best way to fit the data irrespective of a Weibull distribution goodness-of-fit ranking. Smallest extreme value (SEV)

---

3. For example, β-slopes are usually around: 3–4 for corrosion-erosion, and 3–6 for metal fatigue failures.

distributions (associated with the Weibull distribution) can model such things as strength as a characteristic skewed to the left from weak units. (For information, largest extreme value distributions are often skewed to the right as particularly strong units but in a failing sense—representing such phenomenon as cyclones, earthquakes, and floods.)

The mean and standard deviation of a Weibull distribution are not straightforward to calculate.

Similarly, point and interval estimates of Weibull parameters become laborious without computer support.

NOTE: The mean and standard deviation of a Weibull distribution requires calculating Gamma (Γ) functions (or looking up Gamma function tables):

$$\mu = \eta \Gamma\left(\frac{1}{\beta}+1\right) \text{ and } \sigma = \eta \sqrt{\Gamma\left(\frac{2}{\beta}+1\right) - \Gamma\left(\frac{1}{\beta}+1\right)^2}$$

Sometimes Gamma function tables are only provided for values between 0 and 1, since $\Gamma(v) = (v-1)\Gamma(v-1)$.

NOTE: For a sufficiently large sample size, estimates of confidence intervals of the Weibull distribution parameters are:

$$\beta^{\pm} = \hat{\beta} e^{\pm(1.049 z_{\alpha/2})/\sqrt{n}} \text{ and } \eta^{\pm} = \hat{\beta} e^{(\pm 1.018 z_{\alpha/2})/(\hat{\beta}\sqrt{n})}.$$

Distribution-EZ

We can use the Excel analysis tool provided, Distribution-EZ, to both assess how well our data set is represented by the Weibull distribution, and to calculate associated distribution parameters.

The next chapter also describes a way to obtain Weibull parameter estimates using an Excel probability plotting tool. This tool can be particularly valuable if we have censored data.

Applying the Weibull Distribution Model

If we determine that a Weibull distribution provides a reasonable model for the phenomena under investigation, we can either assess whether another distribution (such as exponential, lognormal, or normal distribution) might

Applying the Weibull Distribution Model

better represent the data (based on the Weibull β parameter estimate). We might also proceed to model the phenomena under investigation using the Weibull distribution parameter estimates, and the formula in Table 5.1.

Once we know Weibull distribution parameters, we can easily make calculations using Excel formulas, as shown in Figure 5.2.

Table 5.1
Weibull Distribution Formulas

	Mathematical Formulas	Excel Formulas
PDF $f(t)$	The PDF for the Weibull distribution is rather complex: $f(t) = \frac{\beta}{\eta}\left(\frac{t}{\eta}\right)^{\beta-1} e^{-\left(\frac{t}{\eta}\right)^{\beta}}$	=WEIBULL.DIST(t, β, η, FALSE) or =WEIBULL(t, β, η, FALSE)
CDF $F(t)$	$F(t) = 1 - e^{-\left(\frac{t}{\eta}\right)^{\beta}}$	=WEIBULL.DIST(t, β, η, TRUE) or =WEIBULL(t, β, η, TRUE)
$R(t)$	$R(t) = e^{-\left(\frac{t}{\eta}\right)^{\beta}}$ and the time associated with any particular reliability $R(t)$ is given by*: $t = \eta\{-\ln R(t)\}^{\frac{1}{\beta}}$	=1- WEIBULL.DIST(t, β, η, TRUE)
$h(t)$	The Weibull hazard function is in the form of a power law: $h(t) = \frac{\beta}{\eta^{\beta}}(t)^{\beta-1}$	Use the definition, $h(t) = f(t)/R(t)$

*So the median is given by $T_{50} = \eta\{\ln 2\}^{\frac{1}{\beta}}$.

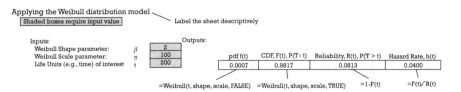

Figure 5.2 Applying the Weibull distribution model using Excel.

CHAPTER 6

Contents

Introduction

Wei-EZ

Choosing the Best Distribution and Assessing the Results

Life Data Analysis: Weibull Probability Plotting

Introduction

To estimate parameters of a probability distribution to model nonrepairable item life, probability plotting is both extremely useful and the least mathematically intensive method. Even with small sample sizes, it yields a graphical picture of how well the distribution fits the data set,[1] and estimates of the distribution parameters. Importantly, probability plotting can handle censored data,[2] whereas some standard statistical formulas (e.g., for the mean and standard deviation) cannot.

---

1. A quantitative measure of goodness-of-fit can also be obtained through measures such as coefficent of determination, also commonly known as R^2 value, where a value of 1 represents a perfect fit and a value of 0 represents no correlation between the data set and a straight line.
2. Censoring may be called for even in situations in which all test items are run to failure. (E.g., when analysis indicates two or more failure modes, we may censor items that have not failed through the mode under study, in order to make use of all our data in describing a particular failure mode.) This requires, of course, that each failed item be examined to determine the failure mode.

Manual probability plotting in general involves plotting the data on specially constructed probability plotting paper (which is different for each statistical distribution). The axes of probability plots are transformed in such a way that the true CDF plots as a straight line.

NOTE: Take, for example, the Weibull distribution. By taking two natural logarithms of the Weibull CDF equation, it takes a linear form of $y = mx + c$:

$$\ln \ln\left(\frac{1}{1-F(t)}\right) = \beta(\ln t) - \beta \ln \eta$$

Therefore, if the plotted data can be fitted by a straight line, the data set fits the selected distribution, and with further constructions we can estimate the distribution parameters. While most of probability plotting these days is done with the use of computer software, understanding probability plotting is useful since many computer software displays and reports are based on this format, so it also provides a good basis for using software tools.[3]

In reliability engineering, the most widely utilized probability plotting distribution used for life data analysis is the Weibull distribution, because it is flexible and the distribution parameters are easy to interpret and relate to the hazard rates and the bathtub curve concept.

After plotting each data point on the Weibull paper or screens, we draw the best fitting straight line through those points.[4] The Weibull distribution has two parameters.[5] The shape parameter β can be determined as the slope of that line (graphically or arithmetically) and scale parameter η can be determined as the time corresponding to 63.2 % of failures (or unreliability).

Significantly, all Weibull CDF curves intersect at a common (relative) point (regardless of the parameter value),[6] at a CDF value of approximately 0.632. The associated t value represents the scale parameter. This characteristic is used in interpreting Weibull plots.

---

[3]. For sample sizes of 50 or greater, MLE of the Weibull distribution parameters generally gives superior results, but for sample sizes of 5–15, simply using Weibull probability paper can give comparable results to MLE.

[4]. A line which gives a good eyeball fit to the plotted data is usually satisfactory, but a simple and accurate procedure to use is to place a transparent ruler on the last point and draw a line through this point such that an equal number of points lie to either side of the line. (Since the plotted data are cumulative, the points at the high cumulative proportion end of the plot are more important than the early points.)

[5]. A three parameter version also exists.

[6]. Consider the mathematical formula for CDF when $t = \eta$.

NOTE: This is a similar property to the exponential distribution though, unlike the exponential distribution, this in general does not represent the MTTF. The Weibull distribution MTTF or expected value is given by the Gamma function:

$$MTTF = E(t) = \eta \Gamma\left(1 + \frac{1}{\beta}\right)$$

However, identifying MTTF without other information can potentially mislead people as representing a constant value throughout any time period. Further, for the Weibull distribution, the MTTF is generally not as useful as the median or the scale parameter (also known as characteristic life).

Wei-EZ

We can perform Weibull analysis using the Excel analysis tool provided, Wei-EZ, to both assess how well our data set is represented by the Weibull distribution, and to calculate associated distribution parameters.

An output of the Wei-EZ plotting application is shown in Figure 6.1.[7]

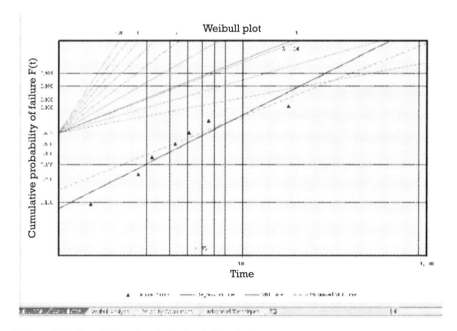

Figure 6.1 Excel Wei EZ probability plotting tool.

---

7. The data set used here is the interarrival data of Figure 2.1 showing a shape parameter of 1.04. This is intended to underscore key points of Chapter 2.

Most probability plotting software applications provide user-friendly interfaces for data input, but users should be aware of two adjustments that such applications make (and that users should make if plotting by hand):

- *Ranking data.* As outlined when discussing the nonparametric cumulative plots, the simplest way to rank each ordered data point, i, is to use rank: $r_i = i/N$, but it is often better to allow for each failure representing a point on a distribution, so median ranking is commonly applied using Benard's approximation (especially in manual probability plotting):[8]

$$\text{Median rank } \hat{F}(x_i) = \frac{i - 0.3}{N + 0.4}$$

- *Handling censored data.* In many cases when (nonrepairable item) life data are analyzed, all of the units in the sample may not have failed or the exact times-to-failure of all the units are not known (i.e., censored data exist). When dealing with censored data, censored items are not plotted as data points on the graph, but their existence affects the ranks of the remaining data points, so the ranks are adjusted to reflect the uncertainty associated with the unknown failure time for the censored items.

NOTE: A common procedure is the rank adjustment method:
- List order number (j) of failed items ($j = 1, 2, ..$);
- List increasing ordered sequence of life values (t_j) of failed items;
- Against each failed item, list the number of items that have survived to a time between that of the previous failure and this failure (or between $t = 0$ and the first failure);
- For each failed item, calculate the mean order number j_{tj} using: $j_{tj} = j_{tj} + jN_{tj}$ where

$$N_{tj} = \frac{(n + 1) - jt_{(j-1)}}{1 + (n - \text{number of preceding terms})}$$

and n is the sample size;
- Then calculate median rank for each failed item, using the median rank approximation:

$$r_j = \frac{j - 0.3}{N + 0.4}$$

---

8. Other methods use the cumulative binomial distribution and its algebraic approximation.

Even though the rank adjustment method is the most widely used method for performing censored items analysis, only the position where the failure occurred is taken into account, and not the exact time. When the number of failures is small and the number of censored items is large and not spread uniformly between failures, this shortfall becomes significant. In such cases, the maximum likelihood method is recommended to estimate the parameters. Further, the Kaplan-Meir method is particularly suited for censored data (including multiple-censored data). The Kaplan-Meir formulation is:

$$\widehat{F(t)} = 1 - \prod_{tj \leq t_1}^{j \varepsilon S} \frac{n-1}{n-j+1}$$

noting that, with uncensored data, $n - j + 1$ is simply the reverse rank from n to 1, and $n - j$ is the reverse rank from $n - 1$ to 0. But when the Kaplan-Meir method is applied to data without censoring, it actually degrades to the rudimentary expression j/n for which the median rank method provides a better estimate.

Choosing the Best Distribution and Assessing the Results

The distribution which provides the best fit to a set of data is not always clear but should consider both:

- How well available distributions fit the data, that is goodness-of-fit.

NOTE: Goodness-of-fit tests include χ^2 (Chi-square) and Kolmogorov-Smirnov (K-S) methods. See Chapter 4 appendix. For probability plotting using the rank regression (least squares) method, the correlation coefficient is a measure of how well the straight line fits the plotted data. For maximum likelihood estimation, the likelihood L, would best characterize its goodness-of-fit.

It is important to note that Weibull plots are to a large extent self-aligning, since succeeding points can only continue upwards and to the right, and axis scales used tend to compress data plots. Therefore, this technique will nearly always indicate good correlation with any best-fit straight line drawn through such points. Understanding the failure rate trend of the lognormal distribution (increasing-decreasing pattern) vs. that of Weibull distribution with $\beta > 1$ (increasing pattern) can help.

- The physical nature of observed failures. Therefore, engineering knowledge and understanding of the item will also be a factor in determining the best distribution.

Example: Battery Life

Ninety-six lithium iron phosphate batteries were put through a charge–discharge life cycle test, using a lithium iron battery life cycle tester with a rated capacity of 1,450 mA h, 3.2V nominal voltage, in accordance with industry rules. A battery sample was deemed to fail when the battery capacity reached 1,100 mA h or less. Table 6.1 provides the number of cycles to failure for the batteries in the sample [1].

A (Distribution EZ generated) histogram of this data set is shown in Figure 6.2, showing the data is left-skewed. Using Wei EZ, at Figure 6.3 we see that a Weibull distribution with a shape parameter of 12.08 and a scale parameter of 890 cycles provides reasonably good fit to the data (i.e., a straight line). Figure 6.4 shows this Weibull distribution overlaid on the histogram data. Therefore, we can use these parameters to model battery life and make predictions (refer to Chapter 5).

We said in Chapter 5 that Weibull shape parameter values beyond 6 indicate very high onset of failures and possibly an effective failure-free period (not normally experienced in most applications). Such a high shape parameter value in this situation might possibly reflect battery capacity degrading progressively, combined with a failure threshold value set by intended battery applications more than any structural (catastrophic) battery failure. Therefore, this data might be a good candidate to refer to a reliability engineer to perform a three-parameter Weibull analysis in order to determine an effective failure-free period.

NOTE: While not intended to replace formal three-parameter Weibull analysis, as an exercise readers might modify the data set with a failure-free period of one cycle less than the first failure (i.e., subtract 509 from all the data points). Doing so yields a Weibull shape parameter value of around 4, indicating that

Table 6.1
Lithium Iron Phosphate Battery Lifetime Data
(in Charge-Discharge Cycles)

510	551	591	622	681	701	707	721	733	740
742	749	751	758	763	774	780	788	792	795
799	803	804	808	811	816	819	821	822	825
826	828	829	830	833	836	837	839	841	844
845	848	849	855	858	860	861	861	864	867
869	874	874	878	879	882	890	892	893	895
898	900	902	902	903	905	906	908	909	912
913	914	914	917	918	923	925	926	928	929
931	932	935	936	939	944	945	948	953	960
961	977	989	994	1001	1021				

Choosing the Best Distribution and Assessing the Results

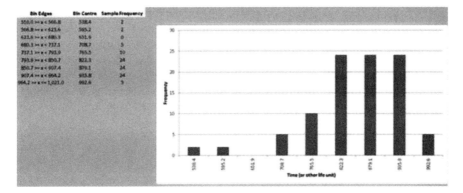

Figure 6.2 Histogram of Table 6.1 data.

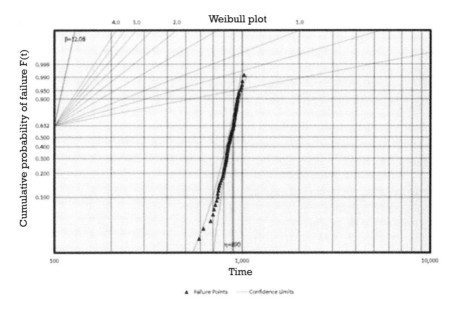

Figure 6.3 Weibull plot of Table 6.1 data (with 95% confidence bounds shown).

failures after the (arbitrary) failure-free period are approximately normally distributed. Ran et al. [1] went on to analyze the physics of the failures reported and suggest a strategy for increasing the life of this battery type by 45.5%.

Example: Jet Engine Failures

Table 6.2 provides failure and service (nonfailure) times (in flight hours) for 31 jet engines at a particular base [3].

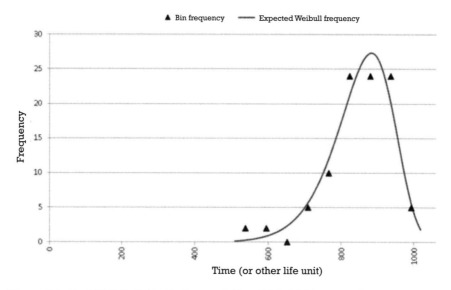

Figure 6.4 Best-fit Weibull distribution overlaid on Table 6.1 histogram data.

A histogram of this data set will not be particularly insightful, with only six failures. Further, if we only used failure data, we would calculate a MTTF of around 772 flight hours. However, we also have another 25 nonfailure data points (22 of which exceed 772 flight hours) that we can incorporate into a Weibull analysis using Wei EZ. One of the strengths of Weibull plotting is the ability to incorporate censored (nonfailure) data.

Figure 6.5 is a Weibull plot incorporating this censored (nonfailure) data, showing a best-fit shape parameter of 1.61 and a scale parameter of 3,559 flight hours. The 95% confidence bounds are relatively large, reflecting the small sample size of failures (and confidence limits of Weibull plots with less than ten failures should not be relied upon). We might expect mechanical components in jet engines to have an increasing hazard rate. We can use these parameters to model jet engine life and make predictions (refer to Chapter 5).

Table 6.2
Jet Engine Failure and Nonfailure Times (Flight Hours)

Failure times	684	701	770	812	821	845			
Service times (nonfailures)	350	850	950	1150	1250	1550	1750	1950	2050
	650	850	1050	1150	1350	1550	1850	2050	
	750	950	1050	1250	1450	1650	1850	2050	

Choosing the Best Distribution and Assessing the Results

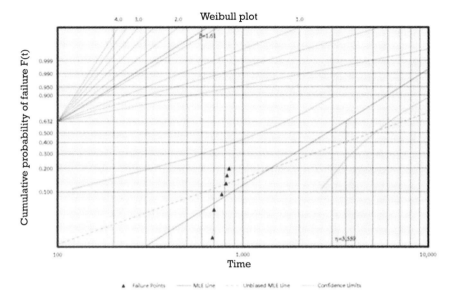

Figure 6.5 Weibull plot of Table 6.1 data (with 95% confidence bounds shown).

Anomalies

Sometimes the failure points do not fall along a straight line on the Weibull plot:

- *Gentle curves.* If the points fall on gentle curves, the origin of the age (time or other life units) scale may not be located at zero (e.g., if the true start time is unknown, one may have been arbitrarily set). Or a true failure-free period may be present (e.g., with roller bearing unbalance failures, the bearing cannot fail instantaneously—it takes many rotations for the wobbling roller to destroy the cage). This problem can be corrected using a more complex (3-parameter) Weibull model, which includes a location or translation parameter (γ). When this location parameter is known, analysts might continue with the more complex model, or simply make this correction in their data and apply the regular (2 parameter) Weibull analysis to the corrected data which may straighten out these points.

- *Concave curves.* Lognormal data on Weibull plots may also appear curved, concave downward.

NOTE: When determining whether a Weibull or a lognormal distribution should be used:

Look at the physical failure mechanism. The Weibull distribution is a competing risk or weakest link model, and applicable to strength of materials, the lognormal distribution is known as a slow fatigue or wear out model.

A general rule is that for considering early failures (e.g., the time associated with the first 1% of failures) the Weibull distribution is more cautious (e.g., earlier times) and the lognormal distribution is more optimistic. (As noted, the Weibull distribution is associated with the smallest extreme value distribution.)

Since it is linked to the normal distribution, one possible way of checking whether a lognormal distribution should be favored over a Weibull distribution is to take the log of the data and observe whether the histogram resembles a normal distribution.

However, sample sizes of, say, at least 20 are often required to discriminate between these distributions; for smaller sample sizes, the standard (two parameter) Weibull distribution is suggested.

- *Corners, doglegs, and s-curves.* Weibull plots showing doglegs, sharp corners, or s-curves might indicate there is more than one failure mode (i.e., population) present. Therefore analysts might consider attempting to classify the failures into different modes (e.g., by examining the broken parts or failure reports). Then, separate Weibull plots are made for each failure mode, with the failure points from the other failure modes treated as censored units.

- Sometimes however the failure points do fall along a straight line on the Weibull plot but may be misinterpreted. For example, a single Weibull plot for a system (usually repairable) with many failure modes will likely show a β close to one, masking infant mortality and wear out modes of component subsystems and parts. (See Chapter 2.)

Example: Disk Drive Failures

Figure 6.6 shows a Weibull plot of the disk drive failure data at Table 3.1. While the plot shows that a (single failure mode) Weibull distribution is not a particularly good fit for the data, it also suggests that analysts might investigate if the first three or four failures might possibly be from a different failure mode or might have resulted from different operating or usage conditions (or both).

Example: Satellite Reliability

Castet and Saleh report on a failure analysis of satellite data [4]. Failure data was collected and analysed for 1,584 earth-orbiting satellites successfully

Choosing the Best Distribution and Assessing the Results 115

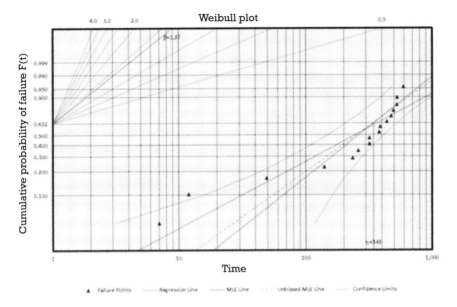

Figure 6.6 Weibull plot of Table 3.1 data (with 95% confidence bounds shown).

launched between January 1990 and October 2008, including culprit subsystems causing satellite failure. Eleven subsystems across 1,584 satellites results in 17,424 (censored and failure) data points. The data used is right censored (i.e., some satellites suffered no failures and some were retired from the sample before failure occurred), with staggered entry times. Both nonparametric (Kaplan-Meir) and parametric (Weibull) analysis was conducted on the satellites and on each of the satellite subsystems, showing remarkably similar results across the two approaches.

Weibull plotting showed well aligned data points, indicating that a Weibull distribution is a good fit to the data (with the exception of the solar array deployment subsystem, a one-shot subsystem). Results indicated that all subsystems except the control processor subsystem (and the solar array deployment subsystem) suffer from infant mortality (with a Weibull shape parameter less than one). This finding has important implications for the space industry and should prompt initiatives to improve subsystem testing and burn-in procedures before launch. The satellite control processor subsystem had a Weibull shape parameter of 1.46, indicating some kind of wear out mechanism occurring. Further, analysis of the battery/cell subsystem suggested two different failure modes—the first around year 3 and the second around year 14.

> **NOTE:** Analysis of the relative contribution of each subsystem to satellite failure showed that the telemetry, tracking, and command (TTC) subsystem was the failure leader up to around year 10 (with relative contribution to satellite failure hovering around 20%) after which gyro subsystem failures began to dominate. Only failures that result in the retirement of the satellite were analyzed.

This example demonstrates the usefulness of reliability data analysis in providing valuable insights to satellite designers and program managers, helping to focus attention and resources to those subsystems with the greatest propensity for and contribution to satellite failure. Note however that no two satellites are exactly alike and every satellite is exposed to different environmental conditions (e.g., different orbits). Therefore, this statistical analysis of satellite failure analyses collective on-orbit failure behavior, working with a relatively large sample in order to obtain a relatively narrow confidence interval for the satellite collective. A disadvantage of this approach is that the reliability calculated may not reflect the specific reliability of a particular satellite. The alternative is to consider each satellite individually, many with a single or no failure, and deriving possibly uncertain specific satellite reliability (i.e., with a wide confidence interval or, as a halfway method, to otherwise specialize the data according to specific platform or mission type).

Exercise

As an exercise, readers might perform a Weibull analysis of the capacitor data in the Chapter 3 appendix to confirm a shape parameter of 5.2, indicating rapid failure onset around the MTTF (1,275 hours) and a characteristic life of 1,378 hours under accelerated conditions.

Further Reading

For readers wanting to explore Weibull analysis in more detail, an excellent place to start is the book by Robert Aberneathy [5].

References

[1] Ran, L., W. Jun-feng, W. Hai-ying, G. Jian-ying, and L. Ge-chen, "Reliability Assessment and Failure Analysis of Lithium Iron Phosphate Batteries," *Information Sciences,* Vol. 259, 2014, pp. 359–368.

[2] Blischke, W. R., and D. N. Prabhakar Murthy, *Reliability Modeling, Prediction, and Optimization,* New York: John Wiley & Sons, 2000, p. 54.

[3] Aberneathy, R. B., J. E. Breneman, C. H. Medlin, G. L. and Reinman, *Weibull Analysis Handbook*, Report No. AFWAL-TR-83-2079, USAF Aero Propulsion Lab, Wright-Patterson AFB Ohio, 1983.

[4] Castet, J., and J. H. Saleh, "Satellite and Satellite Subsystems Reliability: Statistical Data Analysis and Modelling," *Reliability Engineering and System Safety*, Vol. 94, 2009, pp. 1718–1728.

[5] Aberneathy, R. B., *The New Weibull Handbook*, Fifth Edition, Publisher: Robert Abernethy, 2006.

CHAPTER 7

Contents

Memoryless Property

Nonrepairable Items

Repairable Items

MTTF

Distribution-EZ

Point and Interval Estimation for the Exponential Distribution Parameter

Applying the Exponential Distribution Model

Exponential (Continuous) Distribution

The exponential distribution is a continuous distribution, frequently used to model time to failure when the hazard rate is constant (or assumed to be), such as in the useful life portion of the bathtub curve.[1,2] One reason for the popularity of the exponential distribution is its simplicity: it has a single easily-understood parameter to describe the distribution—the constant hazard rate, often denoted by λ. See Figure 7.1.

Memoryless Property

The exponential distribution is sometimes referred to as memoryless since the hazard rate at any time does not depend on how much time has already elapsed and so is not influenced by what has happened in any other time period. For continuous distributions, this characteristic is unique to the exponential distribution.

---

1. Such might be the case for a continuously operating system (using calendar time), or for an intermittently operating system (using operating time).
2. A Poisson process generates a constant hazard rate.

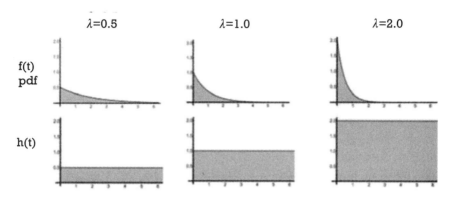

Figure 7.1 Exponential distributions.

Nonrepairable Items

Importantly, with nonrepairable items (e.g., many parts), the constant hazard rate of the exponential distribution implies that no wear out mechanisms occur (fatigue, corrosion, wear, embrittlement, etc.), and no infant mortality occurs. Rather, all failures occur through random extremes in the operating environment (such as power surges, vibration, mechanical impact, temperature fluctuations, moisture variation, nails on roads, etc.), generally occurring randomly throughout the item's life. Accordingly, preventive maintenance does not make sense since the hazard rate before preventive maintenance is the same as the hazard rate after preventive maintenance, and the mean residual life remains unchanged (at $1/\lambda$).

Repairable Items Caution

With repairable systems, a relatively constant ROCOF—no increasing or decreasing trend of cumulative times to failure—implies that, for that individual system, failures can be reasonably assumed to come from the same distribution (since the mean time between failures is constant; refer to Chapter 2.) While this does not necessarily mean that the underlying distribution is exponential, such an assumption is often made when analysts are only concerned with the MTBF of repairable systems (and not the spread), since the exponential distribution is the easiest continuous distribution to work with mathematically. However, note that a constant ROCOF in a repairable system does not imply that no wear out mechanisms occur in the underlying sub-systems and parts. Indeed, digging deeper (analytically) and addressing potential part wear out issues in repairable systems (e.g., through use of more durable parts, or timely part replacement before failure) can substantially reduce ROCOF rates (i.e., increase MTBF). Chapter 2 emphasizes that ROCOF is a different concept to hazard rate, in a similar way that the rate of a person getting sick (and recovering) is different from human mortality rates. Chapter 3 points out that MTBF strictly refers to repairable items and MTTF refers to nonrepairable items but, in practice, MTBF is often used interchangeably to refer to both repairable and nonrepairable items. Because of this loose terminology, it is better to simply consider whether the issue is about repairable systems or nonrepairable items (e.g., parts) rather than the strict definition of MTBF.

MTTF

The mean[3] of an exponential distribution is $1/\lambda$. When only MTTF (i.e., one parameter without reference to a timeframe) is provided, on many specifications for example, an exponential distribution is inherently implied. By definition, MTTF is $1/\lambda$, and replacing the random variable t with this value in the formula for reliability ($R(t) = e^{-\lambda t}$) gives the item's reliability at this time of $R(MTTF) = e^{-1} = 37\%$. So, contrary to popular belief that MTTF shows the time when 50% of units have failed, 63% of units will have failed by the MTTF when a constant hazard (failure) rate applies. This effect is a characteristic of skewed distributions like the exponential distribution.

NOTE: The median of an exponential distribution, where 50% of the population has failed, is calculated as:

$$t = \frac{\ln 0.5}{\lambda} = \frac{0.693}{\lambda}$$

Distribution-EZ

We can use the Excel analysis tool provided, Distribution-EZ, to both assess how well our data set is represented by the exponential distribution, and to calculate the associated distribution parameter.

Point and Interval Estimation for the Exponential Distribution Parameter

We should estimate the parameter and confidence bounds for an exponential model of our data if exploratory data analysis (such as Weibull plotting and analysis) indicates that an exponential distribution is a good fit for the

---

3. The standard deviation of an exponential distribution is the same as the mean, $1/\lambda$.

data, or if we cannot validly assume any other distribution, as might be the case when we simply know that a certain number of failures occurred in a single time interval. The MLE for the parameter of the exponential distribution is:

$$\hat{\lambda} = \frac{\text{number of failures}, d}{\text{total time tested including censored units}}$$

NOTE: In the event of zero failures though, the regular formula provides an unreasonable estimate of a zero failure rate. In these circumstances, a more realistic estimate is

$$\hat{\lambda} = \frac{-\ln \alpha}{\text{total time tested including censored units}}$$

with α often chosen as the 50th percentile (0.5), giving:

$$\hat{\lambda} = \frac{0.6931}{\text{total time tested including censored units}}$$

Confidence bounds (one sided) for $\hat{\lambda}$ are given by the Chi-square (χ^2) distribution with confidence level $1-\alpha$:

$$\lambda \chi^2 (1-\alpha, 2d)/2d \leq \lambda \leq \lambda \chi^2 (\alpha, 2d)/2d$$

for failure censored data or complete data (replacing d with the sample size, n). These bounds can be calculated using the Excel CHIINV function. For time censored data, the upper confidence limit (conservatively) assumes a failure immediately after the censoring time interval (i.e., χ^2 (α, $2(d + 1)/2d$)). All confidence limits apply to both nonreplacement and replacement testing—only the calculation of time-on-test changes. (Chapter 1 discusses censored data.)

NOTE: This uses the relationship between the exponential and Chi-square distributions:

$$\frac{2d\lambda}{\hat{\lambda}}$$

is χ^2 distributed with $2d$ degrees of freedom, where d is the number of failures. A related relationship used extensively in applied statistics pertains to sample and population variance:

$$\frac{(n-1)s^2}{\sigma^2}$$

is χ^2 distributed with (n-1) degrees of freedom.

Example: Aircraft Radar Receiver-Exciter Failures

Table 7.1 provides times (in flight hours) to first failure of aircraft radar receiver-exciters [1–2].

This dataset of 241 failures is interval censored or grouped into 25 equal intervals of 40 flight hours, so a histogram with 25 bins can be easily plotted. We can also expand this table to 241 individual failure data points (using the midpoint of each relevant interval) and use Distribution EZ to fit various distributions to the data. Doing so shows that the best fit Weibull distribution has a shape parameter of 1, indicating that an exponential distribution may model the data well, as Figure 7.2 shows.[4]

Distribution EZ shows that the best-fit exponential distribution mean is 247.2 flight hours, which matches our MLE for the parameter of the exponential distribution of :

$$\hat{\lambda} = \frac{241 \text{ failures}}{59580 \text{ total flying hours} \left(= \text{ the sum of each average failure time}\right)}$$

$$= \frac{1 \, failure}{247.2 \, flight \, hours} = 0.0040449$$

We can use this parameter to model receiver-exciter life and make predictions.

Table 7.1
Radar Receiver-Exciter Failure Data

Hours	Fail	Hours	Fail	Hours	Fail	Hours	Fail	Hours	Fail
0-40	39	200-240	16	400-440	7	600-640	5	800-840	2
40-80	35	240-280	13	440-480	7	640-680	2	840-880	3
80-120	25	280-320	15	480-520	8	680-720	3	880-920	2
120-160	15	320-360	10	520-560	4	720-760	3	920-960	1
160-200	14	360-400	5	560-600	4	760-800	2	960-1000	1

---

4. Wei EZ is only able to handle up to 100 data points so was not able to be utilized in this example.

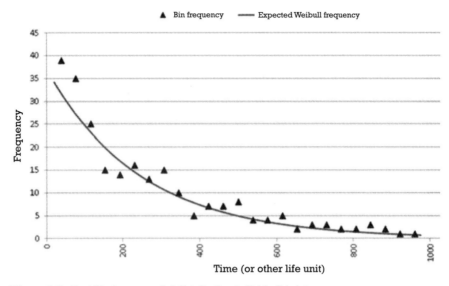

Figure 7.2 Best fit of exponential distribution to Table 7.1 data.

Example: Helicopter Component Failures

Times to failure of three helicopter component types, which together accounted for 37% of system (helicopter) failures are given in Table 7.2 [3].

If we restrict out attention to Part 3 for the time being, we can use Distribution EZ to develop a histogram of this limited data set (see Figure 7.3). While this looks likely to be able to be represented by an exponential distribution, a Weibull analysis showing a shape parameter of 1.2 as the best Weibull fit provides additional support (see Figure 7.4).

Accordingly, we can model the lifetimes of Part 3 using an exponential distribution, with parameter calculated as:

$$\hat{\lambda} = \frac{9 \text{ failures}}{10283.4 \text{ total component operating hours}} = 0.0008751$$

Table 7.2
Helicopter Component Failure Times (in Component Operating Hours)

Part 1	406.3	213.4	265.7	337.7	573.5	744.8	156.5	1023.6	774.8
	644.6	213.3	265.7	337.7	573.5				
Part 2	495.90	207.53	410.10	209.48	750.10	564.5	16.90	410.10	573.60
	117.30	207.53	573.60	270.20	750.10	392.10	920.60	354.50	
Part 3	4057.0	1163.0	607.4	158.7	1088.4	1199.8	420.0	751.1	838.0

Point and Interval Estimation for the Exponential Distribution Parameter 125

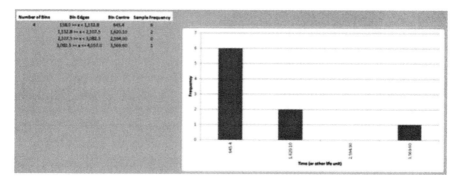

Figure 7.3 Histogram of Table 7.2, Part 3 data.

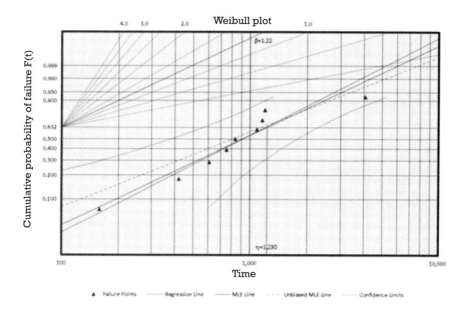

Figure 7.4 Weibull plot of Table 7.2 Part 3 data.

Some readers might legitimately ask why there is no censored (unfailed component) data in this helicopter component failures example, given that all these components (and the remaining 63% of component failures not listed) were presumably operating at the time of other component failures, and that this censored data may alter these results considerably. (Refer back to the jet engine failures example in Chapter 6.) The lack of such information highlights the widespread misconception that censored data is not important in reliability data analysis.

Applying the Exponential Distribution Model

If we determine that an exponential distribution provides a reasonable model for the phenomena under investigation, we can proceed to model the phenomena using the exponential distribution parameter estimates, and the formula at Table 7.3.

We can make point and interval estimates for the exponential distribution parameter using Excel and then make calculations using Excel formulas, as shown in Figure 7.5.

To confirm you are comfortable applying the exponential distribution using Excel, consider the following. To estimate the MTTF of an item type, a life test equivalent to 1,000 hours of operation was conducted. Ten items were tested concurrently in a nonreplacement test and the test was terminated at 1,000 hours (i.e., time censored). One item failed at 450 hours and a second item failed at 850 hours; the remaining 8 items did not fail. What is the total test time? Assuming an exponential distribution is appropriate: What is the best estimate of the item MTTF? What is the lower 90% confidence limit of MTTF? What is the probability of an item surviving for 4,650 hours? For 10,000 hours? If the test was repeated for a life test equivalent of 2,000 hours of operation, how many items would be expected to fail? (Answers: 9,300 hours, 4,650 hours, 2,391 hours, 36.8%, 11.6%, 3.5).

Table 7.3
Exponential Distribution Formulas

	Mathematical Formulas	Excel Formulas
PDF, $f(t)$	$f(t) = \lambda e^{-\lambda t}$ where: λ is the constant rate (e.g., failure rate), and t is time (or some other measure of product use such as cycles, km, etc.).	=EXPONDIST(t, λ, FALSE)
CDF, $F(t)$	$F(t) = 1 - e^{-\lambda t}$	=EXPONDIST(t, λ, TRUE)
$R(t)$	$R(t) = e^{-\lambda t}$	=1-EXPONDIST(t, λ, TRUE)
$h(t)$	The hazard rate of the exponential distribution is a constant value: $h(t) = \lambda$	

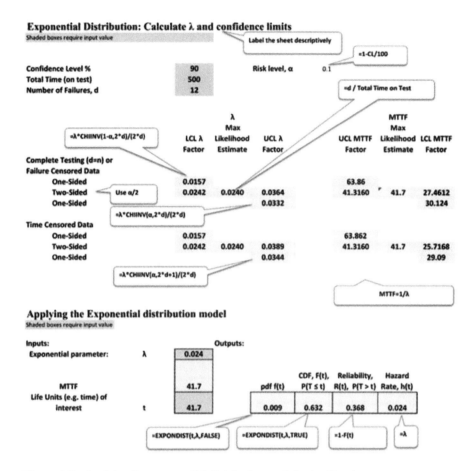

Figure 7.5 Applying the exponential distribution model using Excel.

References

[1] Blischke, W. R., and D. N. Prabhakar Murthy, *Reliability Modeling, Prediction, and Optimization*, New York: John Wiley & Sons, 2000, pp. 51–52 and 60–61.

[2] Lakey, M. J., "Statistical Analysis of Field Data for Aircraft Warranties," *Proceedings of Annual Reliability and Maintainability Symposium*, 1991, pp. 340–341.

[3] Luxhoj, J. T., and H. J. Shyur, "Reliability Curve Fitting for Aging Helicopter Components," *Reliability Engineering and System Safety*, Vol. 38, 1995, pp. 229–234.

CHAPTER 8

Contents

Introduction

Distribution EZ

Applying the Normal Distribution Model

Appendix 8

The Central Limit Theorem

Normal (Continuous) Distribution

Introduction

Though considered the most important distribution in statistics (both theory and practice), the normal distribution (also known as the Gaussian distribution and informally called the bell curve) seldom directly applies in reliability engineering data analysis practice, except for:

- When considering preventive maintenance;
- Both stress and strength measurements which often exhibit normal characteristics.

NOTE: When both stress and strength exhibit normal characteristics, the interaction between stress and strength is also conveniently described by a normal distribution to describe the probability of failure with a mean of

$$u_{difference} = u_{strength} - u_{stress}$$

and a standard deviation of

$$\sigma_{difference} = \sqrt{\sigma^2_{strength} + \sigma^2_{stress}}$$

In general, the critical values of this distribution are when they are negative, equating to the probability of failure (when the stress is greater than the resisting strength). If inadequate, reliability can be improved by further separating the means (reducing mean stress and/or increasing mean strength), or reducing variation of stress and/or strength, or both.

Utilizing the Central Limit Theorem (see the appendix), the normal distribution is also vital in statistical process control (SPC), analysis of variance (ANOVA) and design of experiments—important tools of the reliability engineer's toolbox, but which do not directly relate to the scope of this book.

However, the normal distribution forms the basis for the lognormal distribution and inference testing, both of which are often used in reliability engineering data analysis.

All normal distributions have a characteristic bell curve that extends to ±∞ and, for any given number of standard deviations from the mean, have the same coverage (area under the PDF)—see Figure 8.1. For the normal distribution:

- Values less than one standard deviation away from the mean account for approximately 68% of the set;
- Two standard deviations from the mean account for approximately 95%;

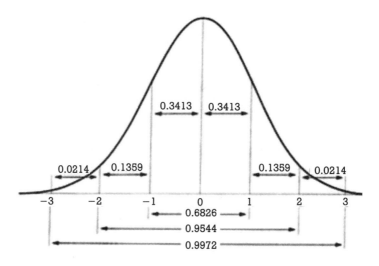

Figure 8.1 Standard normal distribution with standard deviation proportions.

▶ Three standard deviations account for approximately 99.7%.

Given its nature, the normal distribution may not be suitable to model variables that have inherently nonnegative values, or that are strongly skewed (i.e., not symmetrical).[1] Indeed, a disadvantage of the normal distribution for reliability engineering modelling is that its random variable can be negative (i.e., covers the entire real line). This limits its applicability to reliability modeling to instances when the mean is relatively large and the standard deviation relatively small, usually less than a third of the mean.

However, the normal distribution can often represent measured characteristics of products and services that are the result of summing (components, processes, etc.).[2] For example, length, thickness, weight, strength, and time can be the result of a summing process.

Normal distributions are described by two parameters, a mean (μ) and a standard deviation (σ). All normal PDF distributions have a bell shape[3] and can be considered as a version of a standard normal distribution (with a mean of 0 and a standard deviation of 1) whose domain has been stretched by factor σ and height similarly scaled by $1/\sigma$ (so the area under the PDF remains 1), and then translated or shifted by μ. (See Figure 8.2.)

NOTE: Any normal distribution with a given mean, μ, and standard deviation, σ, can be evaluated from the standardized normal distribution (defined as a normal distribution with $\mu = 0$ and $\sigma = 1$) by calculating the standardized normal variate z where

$$z = \frac{x - \mu}{\sigma}$$

Distribution-EZ

We can use the Excel analysis tool provided, Distribution-EZ, to both assess how well our data set is represented by the normal distribution, and to calculate associated distribution parameters.

---

1. Such as the human weight, height, or blood pressure, the time to perform a corrective maintenance/repair action, or the price of a share.
2. In some cases, not only is the sampling distribution of a statistic (e.g., mean) normally distributed (as is expected based on the Central Limit Theorem), so is the population being sampled. The normal distribution is not the distribution for effects that act multiplicatively however. Here the lognormal distribution can come to the fore.
3. But not all bell-shaped distributions are normal distributions.

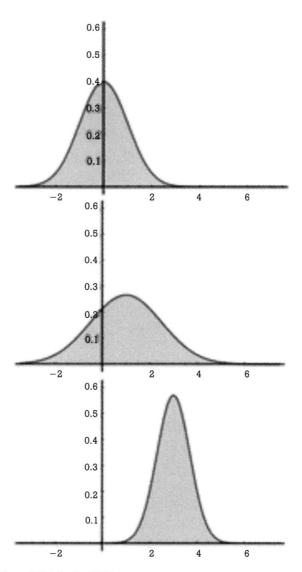

Figure 8.2 Normal distribution PDF.

Point Estimation of Normal Distribution Parameters

The maximum likelihood point estimate for the mean parameter of a normal distribution, and an unbiased estimate of the standard deviation parameter, are calculated the same as for the (nonparametric) mean and standard deviation of any raw data set. The equivalent Excel functions are AVERAGE and STDEV.S respectively.

Confidence Interval for the Mean

The Student's t-distribution can be used to estimate confidence intervals for the normal distribution population mean from sample data (where the population standard deviation is not known and the sample standard deviation is used as an estimate, which is usually the case in reliability engineering). Indeed, the t-distribution was developed for this purpose. In simple terms, a t-distribution might be considered similar to a normal distribution but with wider tails—the relative width of the tails increasing as the sample size decreases (representing greater uncertainty):

- For the desired confidence level determine the appropriate *t*-value based on the sample size, n.[4] Excel has several functions to determine t values, given an area under the t-distribution PDF to represent a given confidence level (e.g., T.INV and T.INV.2T);
- Find the sample mean and standard deviation, $\overline{x}$ and s, as the best estimate for the population mean and standard deviation, $\hat{\mu}$ and $\hat{\sigma}$;
- Calculate the confidence interval as: $\hat{\mu} \pm t\,\hat{\sigma}/\sqrt{n}$

NOTE: For larger sample sizes, the CLT provides a powerful result for all distributions including the normal distribution: the sampling distribution for the mean becomes normal with a standard deviation of

$$\frac{\sigma_{original}}{\sqrt{sample\ size}}$$

Therefore, the confidence interval or margin of error E becomes:

$$\frac{Z_{\alpha/2}\,\sigma}{\sqrt{n}}$$

To determine the sample size required to obtain a given confidence interval or 'margin of error' and confidence level, solve this rearranged formula for n (rounded up to a whole number):

$$n = \left(\frac{Z_{\alpha/2}\,\sigma}{E}\right)^2$$

---

4. Normal distribution sample size requirement estimates. Not surprisingly, the sample size needs to be increased to reduce the confidence interval (sometimes referred to as margin of error) or increase the confidence level.

Usually, 99%, 95%, or 90% confidence intervals are used, with corresponding $Z_{\alpha/2}$ values of 2.575, 1.96 and 1.645 respectively.

When samples are drawn from a normally distributed population, the effect of the CLT is seen more quickly with smaller sample sizes. Nevertheless, the Student's t-distribution provides greater accuracy when the population standard deviation is not known.

Confidence Interval for the Standard Deviation

When the population is normally distributed, the Chi-square distribution is used to estimate the confidence limits for the population standard deviation from sample data:

$$\hat{\sigma}\sqrt{\frac{n-1}{\chi^2_{\alpha/2}}} \text{ and } \hat{\sigma}\sqrt{\frac{n-1}{\chi^2_{1-\alpha/2}}}$$

where χ^2 values use $(n-1)$ degrees of freedom.

Example: Carbon Fiber Failures

Table 8.1 provides failure stresses (in GPa) of single 20-mm carbon fibers [1].

We can use Distribution EZ to show a histogram of this data set at Figure 8.3. While the histogram looks slightly skewed to the left, a normal distribution looks like a promising fit to the data (see Figure 8.4). A Weibull plot (Figure 8.5) confirms the left skew, with a shape factor of 5.54. Accordingly, an analyst might choose to either continue with modeling the data set with a Weibull distribution or with a normal distribution. If we chose to model this data set as a normal distribution (e.g., based on

Table 8.1
20 mm Carbon Fiber Failure Stresses (in GPa)

1.312	1.865	2.027	2.240	2.339	2.478	2.570	2.697	2.821	3.090
1.314	1.944	2.055	2.253	2.369	2.490	2.586	2.726	2.848	3.096
1.479	1.958	2.063	2.270	2.382	2.511	2.629	2.770	2.880	3.128
1.552	1.966	2.098	2.272	2.382	2.514	2.633	2.773	2.954	3.233
1.700	1.997	2.140	2.274	2.426	2.535	2.642	2.900	3.012	3.433
1.803	2.006	2.179	2.301	2.434	2.554	2.648	2.809	3.067	3.585
1.861	2.021	2.224	2.301	2.435	2.566	2.684	2.818	3.084	3.585

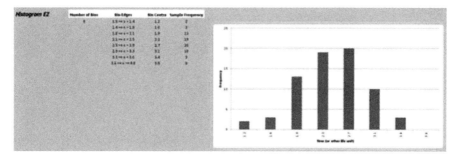

Figure 8.3 Histogram of Table 8.1 data.

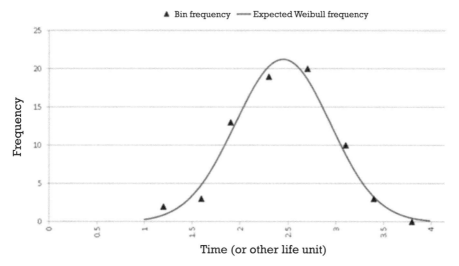

Figure 8.4 Normal distribution fit to histogram of Table 8.1 data.

engineering judgement of the failure mode), we can use the Excel AVERAGE and STDEV.S functions to calculate the relevant distribution parameters of 2.4513 GPa mean and 0.4929 GPa standard deviation respectively.

Applying the Normal Distribution Model

If we determine that a normal distribution provides a reasonable model for the phenomena under investigation, we can proceed to model the phenomena using the normal distribution parameter estimates, and the formula at Table 8.2:

> **NOTE:** Since no explicit (closed form) formula for the CDF of the normal distribution exists, historically, tables using a standardized normal distribution

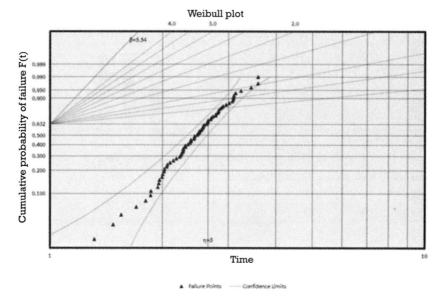

Figure 8.5 Weibull plot of Table 8.1 data.

Table 8.2
Normal Distribution Formulas

	Mathematical Formulas	Excel Formulas
PDF, $f(t)$	$f(x) = \dfrac{1}{\sqrt{2\pi\sigma^2}} e^{-\dfrac{(x-u)^2}{2\sigma^2}}$ where: u is the mean (or expected value) (and also its median and mode); is the standard deviation.	=NORMDIST(x, μ, σ, FALSE)
CDF, $F(t)$	No closed formula exists (i.e., cannot be determined in terms of elementary functions).	=NORMDIST(x, μ, σ, TRUE)
$R(t)$		=1-NORMDIST(x, μ, σ, TRUE)
$h(t)$		Use the definition, $h(t) = f(t)/R(t)$

(defined as a normal distribution with $\mu = 0$ and $\sigma = 1$) were used. Nowadays, spreadsheet functions like =NORMSDIST are used, but old habits die hard and the concept of using the standardized normal variate z and finding the appropriate value of CDF of the z value (represented as $\Phi(z)$) using tables is still used extensively in the literature. Nowadays, the standardized normal CDF, $\Phi(z)$, can also be calculated in Excel as =NORMSDIST(z).

Applying the Normal Distribution Model

Bell [2] provides a (relatively) simple and practical mathematical approximation for the normal distribution CDF, with a maximum absolute error of 0.003:

$$F(t) \approx \frac{1}{2}\left\{1 + sign(t)\left[1 - e^{\frac{-2t^2}{\pi}}\right]^{\frac{1}{2}}\right\}$$

We can make point and interval estimates for the normal distribution parameters using Excel and, once we know normal distribution parameters, also make calculations using Excel formulas, as shown in Figure 8.6.

To confirm you are comfortable applying the normal distribution using Excel, consider the following:

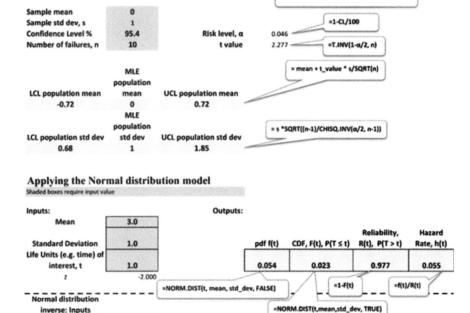

Figure 8.6 Applying the normal distribution model using Excel.

- If data is normally distributed, what is the confidence interval for the mean with 5% level of significance for a sample of ten units with a mean of 15 and a standard deviation of 0.3? (Answer: 14.79, 15.21.)

- The length of a part is normally distributed with a mean of 10 cm and a standard deviation of 0.6 cm. What is the probability that a part will be longer than 10.5 cm? (Answer: 0.202 or 20.2%.)

- The lengths of a part type are normally distributed. How many standard deviation units symmetrical about the mean will include 40% of the lengths? (Answer: +/- 0.52 standard deviations.)

APPENDIX 8

The Central Limit Theorem

The CLT is essentially that:

- The distribution of sample means (of independent random variables, or their sum) is approximately normal even if the population from which the sample was drawn is not normally distributed. The approximation improves as the sample size increases.

- More formally: If the sample size of a random sample is n, then the sample mean has a normal distribution with mean equal to the population mean, and a standard deviation equal to: (population standard deviation)$/\sqrt{n}$. In practice though, if the population is not normal, a sample size of more than four is appropriate and the number of samples should be more than 30.

Note that the CLT alludes to the estimate of the mean being normally distributed with a standard deviation equal to

$$\frac{\sigma}{\sqrt{n}}$$

so if the mean of samples is of particular interest (e.g., in process control), halving the standard error of the sample mean requires sample sizes four times greater. The CLT also supports deriving confidence limits on other (population) parameters, based on sample data.

The CLT can be demonstrated graphically in Figure 8.7 with coin tossing, noting that the normal distribution becomes more pronounced as the number in the sample becomes large:

Based on the CLT, a common practice in many applications is to assume that the variation being analyzed is normal, and so routinely determine

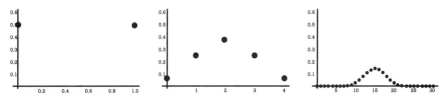

Figure 8.7 CLT applied to coin tossing: probabilities associated with a particular outcome (say, heads) of tossing a single coin 1, 4, and 30 times.

the mean and standard of the normal distribution that best fits the data. However, most variables are curtailed in some way, so assuming a normal distribution may be misleading, especially when used to make inferences well beyond the range of actual measurements.

References

[1] Blischke, W. R., and D. N. Prabhakar Murthy, *Reliability Modeling, Prediction, and Optimization*, New York: John Wiley & Sons, 2000, p. 57.

[2] Bell, J., "A Simple and Pragmatic Approximation to the Normal Cumulative Probability Distribution," 2015, available at SSRN: http://ssrn.com/abstract=2579686.

CHAPTER 9

Contents

Characteristics

Distribution Parameters

Parameter Estimation and Parameter Confidence Limits

Distribution-EZ

Lognormal (Continuous) Distribution

Charactertistics

A lognormal distribution (PDF) is characterized by a single peak (mode), like the normal distribution, but with a long tail (right skewed). Further, it does not have the normal distribution's disadvantage (for reliability analysis) of extending below zero. See Figure 9.1. By definition, if we take the logarithm of a lognormal distributed random variable, it will be normally distributed.[1]

The lognormal distribution is one of the most commonly used distributions in reliability applications (along with the Weibull distribution for initial data analysis), since its right-hand skew makes it suitable for modeling phenomena rates that grow exponentially. It can often model:

- Time to failure characteristics for some electronic and mechanical products (e.g., electrical insulation, transistors, bearings).

---

1. Likewise, if we take the exponential function of a normally distributed random variable, it will be lognormally distributed.

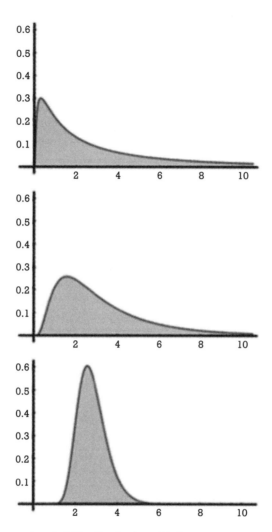

Figure 9.1 Lognormal probability distributions (PDF).

- Fatigue life.
- Usage data such as vehicle mileage per year, and count of switch operations.
- Uncertainties in failure rates.[2]

---

2. Suppose that the best estimate of a variable is $y_{50\%}$ and there is, say, a 90% certainty that this estimate is known within a factor of n (i.e., there is a 90% probability that it lies between $y_{50\%}/n$ and $ny_{50\%}$). This can be described by a lognormal distribution.

- Where the standard deviation of a distribution is a large fraction of the mean, making the use of the normal distribution problematic when random variables cannot logically extend below zero.

- Corrective maintenance (i.e., times to repair an item after failure) for complex systems.

NOTE: There are occasions when the work is performed rather quickly, but relatively unlikely that the work will be done in much less time than usual, and relatively more likely that problems will occur that will cause the work to take much longer. This skews maintenance time distributions to the right and is often well represented by a lognormal distribution. (In addition to the job-to-job variability, there is also variability due to learning, for example if technicians of different experience are being used simultaneously. However, both the mean corrective maintenance time and the spread should reduce with experience and training.)

- Random variables which are the multiplication of other random variables (since their logarithms add). Because many natural growth processes are driven by the accumulation of many small percentage changes, the lognormal distribution describes many natural phenomena, including reference ranges for measurements in healthy individuals (height, weight, blood pressure, etc., after males and female populations are separated), income distributions (for most of the population), exchange rates and share market indices, city sizes, and human behaviors (internet forum comments length, online article dwell time, and length of chess games).

NOTE: If a random variable can be expressed as the sum of component random variables, none of which is dominant, then it can be described by a normal distribution (even if each of the component random variables are not normally distributed or even from the same distribution). A second frequently arising situation is a random variable, say x, that is the product of component random variables. By taking the natural logarithm of the component random variables, we see that $\ln x$ is the sum of the logarithms of the component random variables, and the analogy to the normal distribution becomes clear and x is said to be lognormally distributed.

Example: Solid-State Drives

A solid-state drive (SSD) uses integrated circuit assemblies with no moving mechanical components as memory to store data persistently, as an alternative to hard disk drives and with some significant advantages (typically including lower access time and latency, and less susceptibility to physical shock). A key component of SSDs is flash memory that can store data bits as an electric charge for extended periods of time (e.g., 2–10 years). Typically, the failure governing mechanism for SSDs (and flash memory) is (tunnel oxide) degradation from repeated program (or write) and erase cycles (P/E cycles). A typical single layer cell might endure up to 100,000 P/E cycles and a typical multilayer cell might endure up to 10,000 P/E cycles. A lognormal distribution is often used to describe SSD P/E cycles to failure.

NOTE: A lognormal distribution is often used based on both empirical data and on engineering judgement as follows. When a degradation process undergoes a small increase that is proportional to the current total amount of degradation (i.e., a multiplicative degradation process), then it is reasonable to expect the time to failure (reaching a critical degradation level) to follow a lognormal distribution. This judgement can also apply to other electronic (semiconductor) degradation processes such as corrosion, diffusion, migration, crack growth, and electromigration (and failures resulting from chemical reactions in general). Note however that the lognormal distribution is not always the best model to apply (e.g., Weibull distributions with shape factors below 1.0, indicating some early life failures, have also been used to model SSD failures). See [1–3].

Distribution Parameters

By definition, a random variable follows a lognormal distribution if the natural logarithm of the variable is normally distributed: it is the normal distribution with ln x or ln t as the variate. Indeed, this relationship is so fundamental that the parameters of the lognormal distribution are described in terms of the mean and standard deviation of the logarithm of the variate (i.e., in terms of the transformed normal distribution). Accordingly, some manipulation or transformation (and care) is required when using the lognormal distribution. We will use the subscript τ (Tau) (i.e., μ_τ and σ_τ) to emphasize that the lognormal distribution parameters relate to the mean and standard deviation of the transformed logarithm variate rather than the raw variate.

While this confuses some newcomers, particularly with modern computer applications like Excel performing calculations for us, it might help to consider that in the early days before these applications became widely available, making calculations with the lognormal distribution required transforming the lognormal problem to a normal problem, making calculations based on the normal distribution, and then transforming the results back to the lognormal or real-world space. This knowledge might make the use of μ_τ and σ_τ as lognormal distribution parameters more easily understood.

Note that when $\mu_\tau \gg \sigma_\tau$, the lognormal distribution approximates to the normal distribution (as shown in the third example of Figure 9.1).

Parameter Estimation and Parameter Confidence Limits

If data sets are known to be lognormally distributed, lognormal distribution parameters can be estimated[3] from quantity n sample data, t_1, t_2, ... t_n by:

$$\mu_\tau \approx \overline{t_\tau} = \frac{1}{n}\sum_{i=1}^{n} \ln t_i \; and \; \sigma_\tau \approx s_\tau = \sqrt{\frac{1}{n-1}\sum_{i=1}^{n}\left(\ln t_i - \overline{t_\tau}\right)}$$

That is:

▸ First find the natural logarithm (ln) of each of the given values;

▸ Then calculate the mean and standard deviation of the ln values to determine estimates of the lognormal distribution parameters (e.g., using Excel =AVERAGE(Cell Range) and =STDEV.S(Cell Range)).

Since these (transformed) parameter estimates relate to a normal distribution, confidence intervals for them can be estimated in a similar fashion as for the normal distribution.

Distribution-EZ

We can also use the Excel analysis tool provided, 'Distribution-EZ', to both assess how well our data set is represented by the lognormal distribution, and to calculate associated distribution parameters.

Example: Helicopter Component Failures Revisited

Table 7.2 provided times to failure (in component operating hours) of three helicopter parts. In the case of Part 1, a histogram of the failure data is at Figure 9.2 (using Distribution EZ). While at first glance this histogram might suggest an exponential distribution, the first histogram bin begins at 156 operating hours. While it is conceivable that Part 1 failures have a failure-free period and then undergo exponential failures (beginning with a high number of failures),[4] a distribution with no failure-free period and

---

3. Maximum likelihood estimates.
4. As suggested in the source reference.

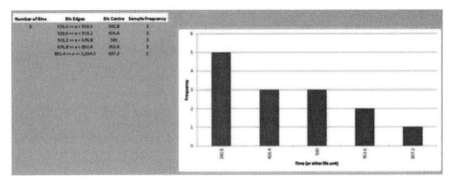

Figure 9.2 Histogram of Table 7.2, Part 1 data.

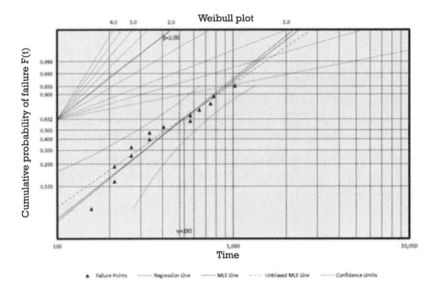

Figure 9.3 Weibull plot of Table 7.2, Part 1 data.

a right skew, such as the lognormal distribution, is more aligned with engineering judgement. Indeed, the Weibull plot at Figure 9.3 supports this view (with a Weibull shape parameter of 2). Accordingly, we can model Part 1 failure data with a lognormal distribution. (Analysis of helicopter Part 2 failure data is left as an exercise for readers.)

Mean and Standard Deviation of Raw Data

Also confusing many newcomers to the lognormal distribution is the calculation of mean and standard deviation. The mean and standard deviation of the Lognormal distribution (in terms of t, rather than $\ln(t)$) is calculated differently from the more traditional methods (e.g., calculating the mean by summing the data values and dividing by the number of values):

$$\mu = e^{\left(\mu_t + \frac{\sigma_t^2}{2}\right)}$$ and then using this value, $\sigma = \mu\sqrt{e^{\sigma_t^2} - 1}$

Correspondingly, when sample data sets are known to be lognormally distributed, the sample mean and standard deviation is calculated as:

$$\bar{t} = e^{\left(\bar{t}_t + \frac{s_t^2}{2}\right)}$$ and then using this value, $s = \bar{t}\sqrt{e^{s_t^2} - 1}$

This different method for calculating mean and standard deviation of a data set (and different results) can sometimes cause confusion. However, the key is whether the sample data set is known to be (or can be reasonably assumed to be) from the skewed lognormal distribution. Being skewed, the mean or expected value for the lognormal distribution is also not the median (T_{50}) or the mode (peak).

NOTE: The median is at e^{μ_t}.

The mean or expected value can be calculated using $T_{50}\, e^{\frac{\sigma^2}{2}}$.

The mode is at $T_{50}\, e^{-\sigma^2}$.

Applying the Lognormal Distribution Model

If we determine that a lognormal distribution provides a reasonable model for the phenomena under investigation, we can proceed to model the phenomena using the transformed normal distribution parameter estimates, and the formula at Table 9.1.

We can make point and interval estimates for the lognormal distribution parameters using Excel and, once we know lognormal distribution parameters, also make calculations using Excel formulas, as shown in Figure 9.4.

To confirm you are comfortable applying the lognormal distribution using Excel, consider the following: Nine items are run to failure with resulting failure times of 240, 180, 143, 42, 57, 120, 71, 100, and 85 days. Determine whether this data set fits well by a lognormal distribution (refer to chapters 3 and 6). If so, determine the probability of a similar item

Table 9.1
Lognormal Distribution Formulas

	Mathematical Formulas	Excel Formulas
PDF, f(t)	$$f(t) = \frac{1}{t}\frac{1}{\sqrt{2\pi\sigma_\tau^2}}e^{-\frac{(\ln t - u_\tau)^2}{2\sigma_\tau^2}}$$ where: is the mean (or expected value) of the transformed logarithm variate is the standard deviation of the transformed logarithm variate.	=LOGNORM.DIST(t, μ_τ, σ_τ, FALSE) or =NORMDIST(LN(t), μ_τ, σ_τ, FALSE)/t
CDF, F(t)	No closed formula exists (i.e., cannot be determined in terms of elementary functions). To apply the translation in reverse, in general use $t = e^{t_\tau}$. To obtain a value corresponding to a particular number of standard deviations (z) or (CDF) probability, use $t = e^{\mu_\tau + z\sigma_\tau}$ where z is the number of standard deviations of the transformed data point from the transformed mean.	=LOGNORMDIST(t, μ_τ, σ_τ, TRUE) To apply the translation in reverse, =LOGINV(probability, μ_τ, σ_τ)
R(t)	No closed formulas exist (i.e., cannot	=1-LOGNORMDIST(t, μ_τ, σ_τ, TRUE)
h(t)	No closed formula can be determined in terms of elementary functions).	Use the definition, $h(t) = f(t)/R(t)$

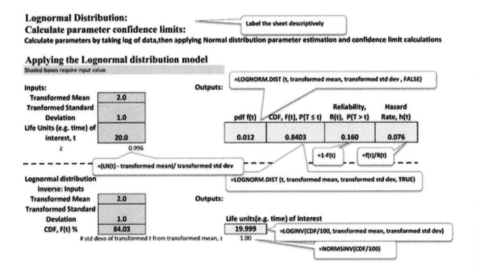

Figure 9.4 Applying the lognormal distribution model using Excel.

failing after the largest recorded failure time of 240 days. (Answer: 0.06 or 6% using nominal values, but as high as 0.29 or 29% with 95% (1-sided) confidence).

References

[1] Kolvogorov, A. N., "On a Logarithmic Normal Distribution Law of the Dimensions of Particles under Pulverization," *Dokl Akad Nauk USSR*, Vol. 31, No. 2, 1941, pp. 99–101.

[2] Sun, F., and A. Parkhomovsky, "Physics-Based Life Distribution and Reliability Modeling of SSD," *IEEE*, 2013.

[3] Sarker, J. and F. Sun, "Reliability Characterization and Modeling of Solid-State Drives," *IEEE*, 2015.

CHAPTER 10

Contents

Introduction

Point and Interval Estimates for Binomial Distribution Parameter

Applying the Binomial Distribution Model

Appendix 10

Combinations

Acceptance (Pass/Fail) Testing

Binomial (Discrete) Distribution

Introduction

Often in reliability engineering, we are interested in only two possible outcomes (for a random variable), for example, failed or operable, reject or accept (see the appendix), heads or tails, > threshold value or ≤ threshold value.

> NOTE: We might convert a measurable variable into a countable one by establishing a threshold value. Similarly, results from a test in which a continuous variable, such as time, could be measured may sometimes be recorded as attributes. For example, several operating units are placed inside an environmental test chamber. At the end of the test each unit is checked to see if it is still operating. The test result recorded for each unit is success or failure. If the unit failed, the exact time of failure is not known.

This binary context is often relevant in acceptance testing/attribute sampling as well as when analyzing the reliability of units that must operate when called on to operate but have no active mission time—these units are referred to as one-shot items.

In such cases, we might be able to calculate simple probabilities in our head (e.g., if 5% of all items produced are defective and we take samples of 20 items, we might expect, on average, that 1 item in each sample (5% of the sample) would be defective).

> **NOTE:** We might intuitively determine that the mean of a binomial distribution is $\mu = np$ where p is the probability or proportion of the population that has failed (or succeeded) and n is the sample size. Less intuitively, the standard deviation of a binomial distribution is
>
> $$\sigma = \sqrt{np(1-p)}.$$

However, probabilities quickly get more difficult to calculate in our heads when considering more complex situations (e.g., the probability of getting three defectives in a sample, or if there are, say, 3 defectives on average and you randomly select 2 items). The binomial distribution models the probability of any number of a particular outcomes in any sample or trial size when the individual probability of two outcomes remains the same for all independent trials.[1]

> **NOTE:** These are formally described as Bernoulli trials. Also note that, since individual probability remains the same for all independent trials, the binomial distribution assumes replacement. Even when this does not occur, the binomial distribution is used as a reasonable approximation for sampling without replacement when the population size is very much larger than each sample. The equivalent distribution which assumes sampling without replacement is the hypergeometric distribution.

The appendix 10 outlines combinations that are the basis of the binomial distribution.

For example, if 25% of a large population has failed, and if 6 parts are randomly selected, the probabilities of 0 through 6 failures in the sample can be calculated separately and shown in Figure 10.1 as a binomial distribution PDF.[2]

Point and Interval Estimates for Binomial Distribution Parameter

The binomial distribution has only two parameters and one of these—the sample size (n)—is often known, leaving only one parameter for statistical

---

1. The binomial distribution can also be used to analyze active redundancy (k-out-of-n) systems.
2. More correctly described as a probability mass function.

Point and Interval Estimates for Binomial Distribution Parameter

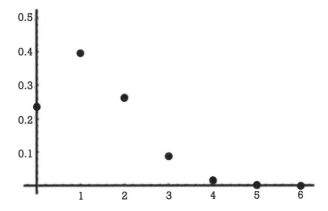

Figure 10.1 Binomial probability plot ($p=0.25$, $n=6$).

analysis (point and interval estimation) from test or sampling data (often in quite restricted amounts).

The (maximum likelihood) estimate for parameter p, the proportion of the population that has failed (or succeeded, depending on your particular definition), is:

$$\hat{p} = \frac{d}{n}$$

where d is the number of failures or defectives found in the sample, and n is the sample size.

Similarly, a reliability estimate (i.e., probability of success or nonfailure) can be obtained from an attribute test.[3] The test could be described as testing n units and recording d failures. The estimate of the reliability of the unit for the conditions of test is

$$\hat{R} = \frac{n-d}{n} \text{ for } d \geq 1.$$

Example: Standby Safety Pump Failures

Low-pressure coolant injections (LPCI) systems are an important safety system in some nuclear reactor types. The LPCI system normally operates in

---

3. A necessary condition for using the binomial distribution as a model in attribute testing is that the probability of success remains the same from trial to trial. This implies that each unit on test has the same probability of success.

standby mode but must be available on demand. LPCI systems include redundant pumps. In a success-fail test of one pump, it failed to start when required in 4 of 240 attempts [1]. The MLE proportion of successes of this pump is 236/240 = 0.9833.

Confidence bounds on p are given by the Beta distribution:

$$Beta_{\frac{\alpha}{2}}(d, n-d+1) \leq p \leq Beta_{1-\frac{\alpha}{2}}(d+1, n-d)$$

where $1-\alpha$ is the confidence (α is the risk). In Excel these bounds are determined by:

$$\text{BETAINV}\left(\frac{\alpha}{2}, d, n-d+1\right) \leq p \leq \text{BETAINV}\left(1-\frac{\alpha}{2}, d+1, n-d\right)$$

NOTE: This confidence bounds formula uses the relationship that if a cumulative form of the unit Beta distribution is repeatedly integrated by parts, it yields a sum of binomial terms for the cumulative binomial distribution. Note the α (alpha) used here to describe confidence bounds should not be confused with the α commonly used as one of the Beta distribution parameters.

Confidence bounds for the binomial distribution parameter can also be found using:

- The F distribution, using the relationship between the Beta distribution and the F distribution. A lower confidence limit for this attribute test reliability estimate is given (using the F distribution) by:

$$R_L = \frac{(n-d)}{(n-d) + (d+1)F_{(\alpha), 2(d+1), 2(n-d)}}$$

where the confidence level is $1-\alpha$. Sometimes this formula is used in reverse to determine the confidence associated with a lower reliability value.
- The binomial distribution itself (using the Clopper-Pearson principle).
- A normal distribution approximation e.g.,

$$\hat{p} \pm z_{\frac{\alpha}{2}} \sqrt{\hat{p}(1-\hat{p})/n}$$

when np and $n(1-p) > 5$.

Zero Failures

When no failures occur in a test however, using the calculation for p to give a claim of 100% reliability may mislead. A lower confidence limit for a zero-failure test is given by:

$$R_L = (1 - \text{confidence level})^{1/n}$$

This equation can be rearranged to solve for the number of units necessary to test without failure to show a given reliability at a given confidence level:

$$n = \frac{\log(1 - \text{confidence level})}{\log R_L}$$

Applying the Binomial Distribution Model

If we determine that a binomial distribution provides a reasonable model for the phenomena under investigation, we can proceed to model the phenomena using the formula at Table 10.1.

Table 10.1
Binomial Distribution Formulas

	Mathematical Formulas	Excel Formulas

PDF, $f(t)$	The binomial PDF* consists of the set of possible x-values and their associated probabilities according to the binomial formula:
	$P(X=x) = {}^nC_x p^x (1-p)^{n-x}$
	where:
	  n is the sample size;
	  p is the proportion of the population that has failed
	  (or succeeded);
	  x is the number of failures (or successes).
	Think of this formula in three parts:
	  The number of possible combinations of failures (without regard to order);
	  The probability of x failures;
	  The probability of $(n-x)$ not failing.
	  (See Appendix on combinations.)
		=BINOMDIST(x, n, p, FALSE)

CDF, $F(t)$	The binomial CDF is the sum of the associated pdf probabilities starting at $x = 0$ up to and including the x-value (i.e. the probability of obtaining x or fewer successes in n trials).	=BINOMDIST(x, n, p, TRUE)
R(t)	Use the definition $R(t) = 1 - F(t)$ and associated formulas	
h(t)	Use the definition, $h(t) = f(t)/R(t)$ and associated formula	

*Strictly, probability mass function.

NOTE: When one-shot devices such as missiles are tested or used, they either perform successfully or not. Various organizations collect pass/fail data over time for one-shot devices. Such data are used to assess reliability and to quantify the uncertainty on reliability when limited amounts of data are collected. A traditional method for analyzing such data is logistic or probit regression with time as the predictor or independent variable. This book does not address logistic or probit regression. Logistic or probit regression suffers from a similar problem as the normal distribution when applied to reliability analysis in that the logistic and normal distributions are defined on the real line, whereas failure times are nonnegative.

An alternative method proposed by Olwell and Sorell [2] is to treat the pass/fail data as right/left censored data respectively and apply the Weibull distribution to this censored data. A fail is treated as a left-censored observation, where the failure occurred before the test, and a pass is treated as a right-censored observation the fail has not (yet) occurred. This method allows one-shot devices to be treated with a traditional failure time analysis using the Weibull distribution.

We can make point and interval estimates for the binomial distribution parameters using Excel and, once we know binomial distribution parameters, also make calculations using Excel formulas, as shown in Figure 10.2.

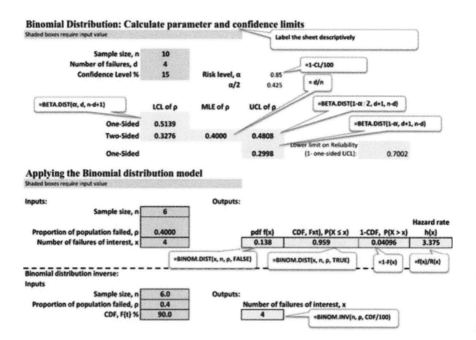

Figure 10.2 Applying the binomial distribution model using Excel.

Applying the Binomial Distribution Model

To confirm you are comfortable applying the binomial distribution using Excel, consider the following:

- The probability of success in a single trial is 0.3 and two trials are performed. What is the probability of at least one success? (Answer: 0.51.)

- A process is producing material that is 30% defective. Five pieces are randomly selected. What is the probability of exactly two good pieces being found in the sample? (Answer: 0.132.)

- A large lot of parts was found by screening to be 20% defective. What is the probability that a sample of 10 parts would have no defectives? (Answer: 0.11.)

APPENDIX 10

Combinations

Much of the difficulty in understanding the binomial distribution relates to combinations. In general, the number of possible ways of selecting or choosing a combination (C) of *r* items from a group of *n*, when the order does not matter, is:

$$^nC_r = \frac{n!}{r!(n-r)!}$$

where ! represents a factorial calculation (eg 3! = 3x2x1=6) and 0! = 1. ^nC_r is read as '(from a sample of) *n* choose *r*' and is also represented as

$$\binom{n}{r}.$$

Most scientific calculators have a factorial key, and often also have a combinations function key. (Calculators have an upper limit to the value of x! If you encounter this, you can use the statistical function in Microsoft Excel which has a larger upper limit.)

In Excel, ^nC_r is calculated using: =COMBIN(*n*,*r*).

For example, if a box of 20 parts has 2 defectives, the number of ways of randomly selecting (choosing) 2 defective parts is $^{20}C_2$ = 20! / (2! x 18!) = 190, so the probability of selecting both defective parts is 1/190.

To check your understanding of combinations you might like to calculate:

- How many ways can eight people be seated in a row of eight seats? (8! = 40 320.)

- How many ways can a committee of four be selected from a group of nine people? ($^9C_4$ = 126.)

- How many possible (5 card) poker hands are there from a 52-card deck? ($^{52}C_5$ = 2 598 960.) What is the chance of getting a royal flush (ace, king, queen, jack, ten all in the same suit)? (With four suits there are four chances in all the possible poker hands or 1/649 740.)

- An organization has six identical generators in stock, two of which are known to be defective:
    - How many different groups of three generators can be selected from the stock? ($^6C_3$ = 126.)

- How many different groups of three generators can be selected that do not include the defective generators? ($^4C_3 = 4$ since we exclude the defective generators.)
- How many different groups of three generators can be selected that contain exactly one defective item? (12. We have two possibilities: If we include only one particular defective generator we have 6 possibilities for the remaining two generators, and if we include only the other defective generator we have another 6.)
- How many different groups of three generators can be selected that contain at least one defective item? (16. This can be calculated as the total possibilities less the number of groups with zero defective generators, $^6C_3 - {^4C_3}$. Alternatively, we can add the number of groups with exactly one defective generator, 12, to the number of groups with two defective generators which is 4 since there are 4 nondefective generators that could be selected for the third generator, $^4C_3 = 4$. Note the binomial distribution is not used here because the population is small and replacement does not occur.)

Sometimes we may have difficulty deciding whether permutations or combinations is required for a particular analysis. The key lies in whether or not the arrangement or order of the items is important.

Acceptance (Pass/Fail) Testing

The binomial distribution can also be used in planning pass/fail type tests (i.e., where there is a unique pass-fail criterion), including acceptance or qualification tests, to provide a degree of assurance (to the buyer or consumer) that no more than some specified fraction of a batch of products is defective. Such tests are used when the cost of testing or inspecting all units is prohibitive or the tests are destructive (or at least damaging). An acceptance testing procedure is typically set up in the following way.

Since only a finite sample, n, is to be tested, there will be some risk that the population will be accepted despite the consumer's criteria or specification that no more than a particular fraction of the total batch fail the test. This is referred to as the consumer's risk (β). In turn, the supplier of the product may believe that their product exceeds the consumer's specification (i.e., possesses a lower actual failure fraction). In taking only a finite sample however, they run the risk that a poor sample will result in the batch being rejected. This is referred to as the producer's risk (α).

The object is to construct a sampling scheme in which the specified and actual defective fractions result in predetermined, balanced (i.e., roughly equal) values of α and β using the relevant CDF formulas with corresponding values of specified and actual defective fractions (i.e., calculations of both should give roughly equal values):

β =BINOMDIST(x, n, ρ_{spec}, TRUE) and α =1-BINOMDIST(x, n, ρ_{actual}, TRUE)

The theoretical procedure is to first consider the first formula as the sample size varies. For a given value of x, and ρ_{spec} and β, a range of sampling schemes can be determined for different values of sample size, n. From the range of sampling schemes (different values of n), one is selected such that the value of α is (approximately) equal to β for a given value of x, and ρ_{actual}. If the sample size to meet this balanced criteria is considered too large, then sometimes acceptable values of risk (both α and β) might be reconsidered. (To reduce both α and β, the sample size needs to increase.)

In practice, several standards provide a range of acceptance or qualification tests using this approach. While the original military standard (MIL-STD-105) is no longer supported, many comparable standards exist, including ANS/ASQ Z1.4 and ISO 2859-1: 1999.[4] A more recent acceptance sampling standard is ISO 28598-1:2017. Other reliability testing standards provide conceptually similar standard test plans based on the Poisson distribution (e.g., for reliability tests). (Often the ρ_{actual} values are stated in terms of multiples of the ρ_{spec} value.)

More varied and sophisticated schemes include double sampling and sequential sampling. Double sampling accepts a very good batch or rejects a very bad batch, but if too much uncertainty exists about the quality of the batch (i.e., for borderline cases), then a second sample is drawn and a decision made based on the combined sample size ($n_1 + n_2$). Sequential sampling extends this idea. The sample is built up item by item and a decision made after each observation to either accept, reject, or take a larger sample. The advantage of this approach is that very good or very bad batches can be accepted or rejected with very small sample sizes, but when more doubt exists then larger sample sizes can be used. A disadvantage is that test planning becomes more complex.

---

4. MIL-STD-1916 (1996), which replaced MIL-STD-105E, and its companion MIL-HDBK-1916 (1999) sought to replace prescribed sampling requirements by encouraging contractors to submit efficient and effective process (i.e., prevention) controls through a comprehensive quality system, a statistical process control program (mentioned in Chapter 8) and continuous improvement.

References

[1] Blischke, W. R., and D. N. Prabhakar Murthy, *Reliability Modeling, Prediction, and Optimization*, New York: John Wiley & Sons, 2000, p. 54.

[2] Olwell, D. H., and A. A. Sorell, "Warranty Calculations for Missiles with Only Current-Status Data, Using Bayesian Methods," *Proceedings Annual Reliability and Maintainability Symposium*, 2001, pp. 133–138.

CHAPTER 11

Contents

Introduction

Markov Models

Point and Interval Estimates for Poisson Distribution Parameter μ

Applying the Poisson Distribution Model

Poisson (Discrete) Distribution

Introduction

Besides the binomial distribution, the other discrete distribution often used in reliability engineering is the Poisson distribution. One reason for the popularity of the Poisson distribution is its simplicity: it has a single easily understood parameter to describe the distribution.

Situations in which discrete events occur randomly over time (or other physical dimensions)[1] often include failures occurring in repairable systems, aircraft accidents, the number of network failures each day, and the number of engine shutdowns each 100,000 flying hours. Such events (with one of two possible countable outcomes) are Poisson-distributed if they occur at a constant average or expected rate, μ.[2] Therefore, you should not be surprised to find that mean of a Poisson distribution is this

---

1. These situations are formally called stochastic point processes.
2. Think of this as the expected number of random events in a region is proportional to the size of the region or time period. More formally, this situation is described by a HPP, which describes a sequence of independently and identically exponentially distributed random variables.

constant average or expected rate.[3] Given that events occur with a known constant average rate and independently of the time since the last event, the Poisson distribution is used to find the probability that an event will occur a specified number of times in a fixed interval of time (or distance, area, volume, etc., depending on the context).

Other common examples of Poisson processes might include the number of emails received each day, vehicle traffic flow past a point, the number of phone calls received by a call center each hour, the number of work-related accidents over a given period of time, the number of goals scored each game in a professional soccer league, and the number of overflow floods in 100 years in a particular region.

> **NOTE:** An intuitively strange characteristic of Poisson processes is sometimes referred to as the waiting line paradox where, for example, you might expect that if buses are coming every 10 minutes and you arrive at a random time, your average wait would be something like five minutes. However, if bus arrivals do follow a Poisson process, then your average wait time will be 10 minutes. A Poisson process is a memoryless process that assumes the probability of an arrival is entirely independent of the time since the previous arrival.

Examples of processes which are not Poisson might include the number of earthquakes in a particular region each decade if one earthquake increases the probability of others (e.g., aftershocks), or processes where zero events occurring in a period are not reasonable (e.g., the number of nights stayed at a particular hotel in a particular month).

Some examples of Poisson distributions are at Figure 11.1.[4] The Poisson distribution is related to many other topics previously discussed in this book, including the HPP outlined in Chapter 2, the exponential distribution, the binomial distribution, and the normal distribution.

However, our minds should turn specifically to the Poisson distribution when we are concerned with countable (discrete) events that occur at a constant average rate.

> **NOTE:** The Poisson distribution can be used to approximate the binomial distribution when the (binomial) sample size, n, is very large (approximating infinity) and the number of defectives, p (or $1 - p$), is relatively small or rare (making calculations easier before computers became widely available, sometimes termed the law of rare events). Indeed, the Poisson distribution can be shown

---

3. The standard deviation of a Poisson distribution is $\sqrt{\mu}$.
4. Another intuitively strange characteristic of the Poisson distribution is that if we reframe the problem so that the interval of time (or space, etc.) of interest is such that the constant average rate of arrival is one, then any Poisson problem can be examined using a single Poisson distribution (where $m = 1$).

Introduction

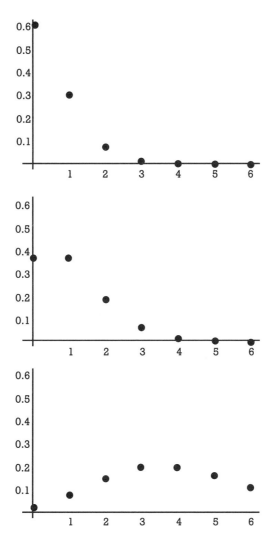

Figure 11.1 Poisson distributions (μ is 0.5, 1, and 4, respectively).

to result from taking the limit of the binomial distribution as p approaches zero and n approaches infinity. A rule of thumb is that the binomial distribution approaches a Poisson distribution with $\mu = np$ when: $n \geq 20$ and $p \leq 0.05$ or if $n \geq 100$ and $np \leq 10$.

The normal distribution represents a limiting case for the Poisson distribution (as well as binomial distribution). When $\mu \geq 10$ the Poisson distribution approaches a normal distribution with mean of μ and standard deviation of $\sqrt{\mu}$. Or use:

$$z = \frac{x + 0.5 - \mu}{\sqrt{\mu}}$$

When $\mu < 10$, the Poisson distribution is skewed to the right.

The Poisson distribution can also be used in reverse (e.g., for spares assessing). The required number of corrective maintenance spares necessary for a given period of time depends on the expected number of failures, which depends on the unit operating time and the failure rate. If the item has a constant failure rate expectation, the probability of requiring no more than a specified number of replacement units can be found using the cumulative Poisson distribution.

NOTE: Using the Poisson distribution for spares assessing assumes that the MTTR is much smaller than the MTTF. Further, since the Poisson distribution considers an asymptotic steady-state process, the MTTR and MTTF are usually sufficient representations of the maintenance and failure distributions. Accordingly, constant repair and failure rates associated with exponential distributions are often used in these calculations, but this should not infer that the underlying distributions are in fact exponential. Note also that the total number of spares kept in inventory in practice is usually more complex since it also depends on the delivery time, the cost of maintaining the inventory, and the availability requirement.

Example: Substation Transformer Spares

Catastrophic failure of a substation transformer can cause energy supply interruption for many consumers (with subsequent loss of revenue, goodwill and market share and possible regulator fines). Given their importance, transformers installed in a substation may include redundancy, so that the substation can continue to meet peak load requirements even if one transformer fails. This is safe, but expensive—a substation system nominally comprising 130 transformers would require double this number for full dual-redundancy, and even 26 additional transformers would be required where one redundant transformer was added for each five-transformer group (and some spares may still be required in each case). Further, even when redundancy is applied initially, due to load growth this arrangement does not always persist. Consequently, many utilities opt for shared spare transformers to avoid significant capital, operations and maintenance expenditure and still provide adequate service reliability. Therefore, the number of spares held for cases of catastrophic failures is an important decision, particularly given that substation transformer rebuilding or procurement time is 1.0–1.5 years (in the case of a catastrophic failure).

Leite da Silva et al. [1] use the Poisson distribution to calculate the optimal number of transformer spares for a utility with 132 transformers (of 72–25 kV, 16 MVA) with individual failure rate of 0.011 failures per year and mean time of one year for repairing/procuring a new transformer (which provides the time frame of interest). Considering a Poisson model reflecting these values with, for example, five spare transformers, the reliability of this system is assessed to be 99.62% and the risk of failure within a one-year period is 0.38%. This value represents the probability of having more than five transformers failed in this period. If this business risk is still unacceptable, we can redo the calculation varying the number of spare transformers, but noting that as each spare is added, the reliability increases in ever smaller increments but the acquisition and inventory costs increase linearly.

Markov Models

Using a technique called Markov modeling, Leite da Silva et al.[1] also examine the possibility of replenishing the inventory through procuring one or more transformers each period. Markov modeling is a powerful technique which extends the use of the Poisson distribution and the HPP to model the transition between two or more possible system states (such as operational state and an under repair state). Markov models [1] are outside the scope of this book but interested readers might refer to Boyd [2] for a good introduction, and Pukite and Pukite [3].

NOTE: Leite da Silva et al. [1] also point out that the reliability of a system composed with field transformers and spares can also be assessed by Monte Carlo simulation. Indeed, to model interactions of many failure distributions and distribution types, Monte Carlo simulation sometimes becomes the only practical option. Monte Carlo simulation can also account for many additional characteristics related with equipment operating memory, time correlation, or physical constraints. Chapter 13 briefly outlines Monte Carlo simulation.

Example: Determining Construction Project Contingency

One of the more common methods of budgeting for contingency in construction projects is to consider a percent of estimated cost, based on previous experience with similar projects. (A more detailed approach is to assign various percent contingencies to various parts of the budget.) Touran [4] developed a novel contingency model which considers the random nature of change orders and their impact on project cost and schedule, assuming that events causing delays and cost adjustments (change orders) occur randomly according to a Poisson process. The approach proposed depends on only a few variables, including an estimate for the average rate of change (such as one change order per month), project duration (such as 24 months), and average cost per change (such as 1% of total cost). The results of the model compared well when compared to a database of 1,576 completed construction projects (e.g., 64% of cost overruns beyond contingency levels predicted, versus 69% actual).

Point and Interval Estimates for Poisson Distribution Parameter μ

The MLE for the Poisson distribution parameter is simply the sample mean for that sample interval, expressed, for example, as either:

$$\hat{\mu} = \frac{\sum_{i=1}^{n} x_i}{n}$$

where n is the sample size; or when exact times or positions are not known:

$$\hat{\mu} = \frac{d}{t}$$

where d is the number of occurrences (e.g., defects or failures) occurring over time t.

(Two-sided) confidence bounds are given by the distribution:

$$\frac{1}{2n}\chi^2_{2*\text{expected events}}\left(1-\frac{\alpha}{2}\right) \leq \mu \leq \frac{1}{2n}\chi^2_{2(\text{expected events}+1)}\left(\frac{\alpha}{2}\right)$$

where $1 - \alpha$ is the confidence (α is the risk). In Excel, these bounds are determined by:

CHISQ.INV $(1 - \frac{\alpha}{2}, 2*$ expected events$)/(2*n) \leq \mu$, and

$\mu \leq$ CHISQ.INV$((\frac{\alpha}{2}, 2*$ (expected events $+ 1))/(2*n)$.

NOTE: Confidence bounds for the Poisson distribution parameter can also be found using a normal distribution approximation e.g.,

$$\hat{\mu} \pm z_{1-\frac{\alpha}{2}} \sqrt{\hat{\mu}/n}$$

Applying the Poisson Distribution Model

If we determine that a Poisson distribution provides a reasonable model for the phenomena under investigation, we can proceed to model the phenomena using the formula at Table 11.1.

We can make point and interval estimates for the Poisson distribution parameters using Excel and, once we know Poisson distribution parameters, also make calculations using Excel formulas, as shown in Figure 11.2.

Table 11.1
Poisson Distribution Formulas

	Mathematical Formulas	**Excel Formulas**
PDF	The probability of x failures is $$f(x) = \frac{\mu^x}{x!} e^{-\mu}$$ where x is a positive integer (0, 1, 2, etc.) and μ is the (constant) mean event expectation for the interval of interest. Where this rate is not given in terms of the specific period of interest, we can use $\mu = \lambda t$ where λ is considered as a (constant) intensity (in time, length, etc.). Since x can take any whole number, the probability distribution technically extends indefinitely.	= POISSON(x, μ, FALSE)
CDF	The CDF, $F(x)$, for the Poisson distribution can be calculated by adding discrete PDF probabilities (this can become tedious by hand).	=POISSON(x, μ, TRUE)
R(t)	Use $R(x) = 1 - F(x)$.	
h(t)	Use the definition, $h(t) = f(t)/R(t)$.	

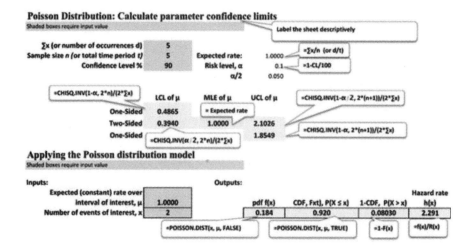

Figure 11.2 Applying the Poisson distribution model using Excel.

To confirm you are comfortable applying the Poisson distribution using Excel, consider the following. An item that operates an average of 2,000 hours per year has a constant MTTF of 10,000 hrs. How many spares will be necessary over a five-year period for corrective maintenance, to provide a 90% probability of having a spare available? (Answer: 2 provides 92% probability.)

References

[1] Leite da Silva, A. M., J. G. de Carvalho Costa, A. A. Chowdhury, "Probabilistic Methodologies for Determining the Optimal Number of Substation Spare Transformers," *IEEE Transactions on Power Systems*, Vol. 25, No. 1, 2010, pp. 68–77.

[2] Boyd, M. A., *An Introduction to Markov Modeling: Concepts and Uses*, NASA Technical Reports Server, Document ID 20020050518, 1998.

[3] Pukite, J., and P. Pukite, *Modeling for Reliability Analysis: Markov Modeling for Reliability, Maintainability, Safety, and Supportability Analyses of Complex Systems*, Piscataway, NJ: IEEE Press, 1998.

[4] Touran, A., "Calculation of Contingency in Construction Projects," *IEEE Transactions on Engineering Management*, Vol. 50, No. 2, 2003, pp. 135–140.

CHAPTER 12

Contents

Introduction

Nondestructive Degradation Sampling Using the Same Samples

Destructive Degradation Sampling (or Using Different Samples), Utilizing a Common Distribution Parameter

Destructive Degradation Sampling, with No Common Distribution Parameter

Analyzing Degradation Data

Introduction

Performance indicators for some kinds of items degrade with time (or other life units) and the item is considered to have failed if the performance indicator passes a specified threshold. If this degradation path can be predicted, we can estimate the item failure time, to help people decide whether to fix or replace the item before its failure. This is the essence of prognostics and health management (PHM), where the health of an item can be monitored, and its remaining useful life estimated (see Chapter 13).

Item degradation data are random variables arranged in order, so can be treated using time series data methods (usually assuming the same failure mechanism applying throughout monotonous, and therefore non-stationary, degradation). Time series data can often be thought of as consisting of a (deterministic) trend component (or components) and a (stationary) random component.

As previously discussed, when choosing the most appropriate model for a data set, we should be considering not only the goodness-of-fit, but also past experience and engineering judgment, and the nature of the data. To select an appropriate trend model, consider the physics of degradation as well as the (empirical or statistical) goodness-of-fit of a variety of likely suspect trend models. The physics of degradation of many items has been well studied and associated degradation models can be found in published papers and books. Such models are often based on common functions, and if specific degradation process models are not available then statistical models based on these common functions can be used to determine a statistical model that best fits the data set. Simple degradation models include:

- Linear: $D(t) = a \times t + b$;
- Logarithmic: $D(t) = a \times \ln(t) + b$;
- Exponential: $D(t) = a \times e^{bt}$;
- Power: $D(t) = a \times t^b$.

Nondestructive Degradation Sampling Using the Same Samples

When we are able to retest the same test item (sample) at various times, a straightforward approach to estimate the reliability is to apply the appropriate degradation trend model/s to each test item, each with its own best-fit coefficient values, to predict the time it reaches the degradation threshold value. Then, analyze the extrapolated failure times in the same manner as conventional life data analysis, using a stationary continuous model (e.g., Weibull, normal, lognormal, and exponential distribution)[1]. Figure 12.1 illustrates the reliability PDF at the threshold, inverted to help reduce clutter. This approach might apply to brake pad wear, crack size propagation, battery voltage degradation, or bulb luminous flux for example. As with conventional life data analysis, the amount of certainty in the results is directly related to the number of samples being tested.

Example: Accelerometer Degradation

The performance of a type of accelerometer used in aerospace and weapons applications degrades over time. When a particular performance parameter reaches a threshold for degraded performance, the accelerometer is deemed to have failed. Table 12.1 (from [1]) shows a subset of the total data

---

1. Other more complex approaches might regard the coefficient values as random variables (e.g., with a bivariate normal distribution).

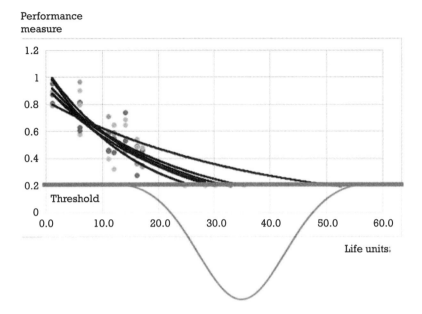

Figure 12.1 Nondestructive degradation analysis.

Table 12.1
Accelerometer Degradation Data

Item	0 Hours	720 Hours	1,440 Hours	2,160 Hours	6,480 Hours	7,200 Hours
1	26.5714	26.5821	26.5572	26.5456	26.4734	26.4638
2	26.5514	26.5434	26.5354	26.5273	26.4791	26.4710
3	26.5388	26.5333	26.5278	26.5223	26.4893	26.4838
4	26.5537	26.5437	26.5336	26.5236	26.4635	26.4535
5	26.5634	26.5517	26.5400	26.5283	26.4581	26.4464
6	26.5405	26.5284	26.5163	26.5042	26.4314	26.4193
7	26.5380	26.5254	26.5128	26.5002	26.4247	26.4121
8	26.5205	26.5069	26.4934	26.4798	26.3986	26.3850
9	26.5462	26.5289	26.5115	26.4941	26.3898	26.3724
10	26.5791	26.5579	26.5367	26.5156	26.3885	26.3673

collected. Given that failure is defined as the performance parameter reaching a value of 25.0, we can extrapolate this performance data and make conclusions regarding failure times and a failure distribution.

We can use Excel to show a scatter plot of the data associated with each item and determine that both linear trend lines and exponential trend lines provide very good fits to each data set. Looking at the linear trend alternative first, using Excel we insert 10 linear trend lines and display the equation for each on the chart. For example, with Item 1 we get a linear trend line of $y = 26.5815 - 0.00002t$. Solving for t when $y = 25$, we obtain 79,075 hours (i.e., by calculating $(26.5815-25)/0.0002$), or 9.03 years (by dividing this result by 8,760 hours/year). When we repeat this process for the other nine samples tested and plot the predicted failure times (e.g., using the Distribution EZ tool), we see that, even with just 10 data points, there is a definite right skew. Therefore, we apply a lognormal distribution to these predicted results to obtain lognormal distribution parameters of $\mu_\tau = 2.7306$ and $\sigma_\tau = 0.4189$, and use this information to calculate an MTBF of 11.7 years.

Turning our attention to fitting exponential trend lines using the Chapter 2 appendix on least squares curve fitting to derive better estimates of parameters, for item 1 we get a least squares fit of an exponential trend line of $y = 26.581e^{-6.19E-07t}$ Solving for t when $y = 25$ (i.e., $\ln(26.581/25)/0.000000617$) we obtain 11.31 years. When we repeat this process for the other nine samples tested and plot the predicted failure times (e.g., using the Distribution EZ tool), we see that, even with just 10 data points, there is also a definite right skew. Therefore, we apply a lognormal distribution to these predicted results to obtain lognormal distribution parameters of $\mu_\tau = 2.4319$ and $\sigma_\tau = 0.3535$, and use this information to calculate an MTBF of 12.1 years [1].[2]

Destructive Degradation Sampling Using a Common Distribution Parameter

In other cases, only destructive measurements of degradation might be possible, such as for adhesive bond strength. When we are not able to retest the same item (sample) at various times (e.g., because testing or measurement is destructive) then we cannot build a degradation model for each individual item. Therefore, we need to regard the degradation value as a random variable with a particular distribution at a given time. Accordingly, the first step of destructive degradation analysis involves using a statistical distribution (e.g., Weibull, exponential, normal, or lognormal) to represent the variability of a degradation measurement at a given time. For example, we might determine that our degradation data set follows a Weibull, normal, or lognormal distribution at any given time.

---

2. Even without the full data set, these exponential trend line results are the same (to 3 significant figures) as in [1].

If we can determine that one or more parameters of our distributions (such as the Weibull shape parameter) is reasonably constant over time [2],then the mathematical modeling becomes more straightforward.

A degradation model combining a deterministic trend component and a stationary random component can be used to represent the degradation, for example:

❱ *Weibull*: If the shape parameter remains reasonably constant, set ln(scale parameter) as a function of time;

❱ *Exponential*: Set ln (MTTF) as a function of time;

❱ *Normal*: If standard deviation remains reasonably constant, set the mean as a function of time;

❱ *Lognormal*: If σ_τ remains reasonably constant, set μ_τ as a function of time.

For example, Figure 12.2 represents the case where a normal distribution with reasonably constant standard deviation is used to represent the stationary component at each time period, and the degradation follows an exponential model.

With this information, we can develop an appropriate distribution for failure at the degradation threshold level. We do so by first determining where the deterministic trend parameter (in Figure 12.2, an exponential trend of the means) meets the threshold value. Then, for each of our common stationary components (in Figure 12.2, each of our normal distribution fits) we determine the probability of failing (exceeding the threshold value), and then fit an appropriate distribution to these threshold probabilities (e.g., using a least-squares fit in Solver) (see Chapter 2 appendix).

NOTE: Accelerated life testing often follows a similar approach. Nelson [2] provides several tips when planning accelerated life tests; those that might apply to more straightforward degradation test planning described here are:

- To reduce statistical uncertainty in the estimates, consider testing relatively more items at the longer test times than at intermediate times.
- Conduct analysis both with and without any outlier data, to see if outliers affect the results appreciably.
- Plot residuals of the plotted data (i.e., the difference between observed and the fitted data). Plots of residuals can be more informative than formal tests for goodness-of-fit, variance, and assumed distribution. Noting that semiconductor devices failure times often follow a lognormal distribution, if the lognormal distribution describes the (stationary) random degradation component then log residuals should look like a sample from a normal distribution.

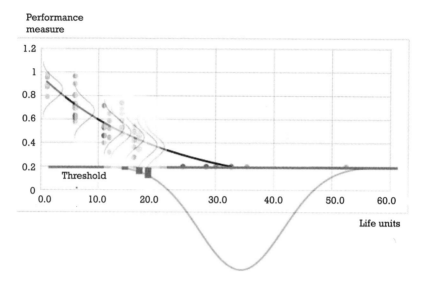

Figure 12.2 Destructive degradation analysis (preferred option).

Destructive Degradation Sampling with No Common Distribution Parameter

When we find, however, that none of the parameters of our distributions (such as the Weibull shape parameter) is reasonably constant over time, then this approach is unsatisfactory and we need to step back from our desire for a model combining a deterministic trend component and a common stationary random component. If we can only reasonably assume that each stationary random component is of the same distribution type, representing the same degradation mechanisms at play but without sharing common distribution parameters, we can simply use each distribution fit (of the same type of distribution) at each period to determine the probability of failing (i.e., exceeding the threshold value), and then to fit an appropriate distribution to these threshold probabilities. That is, we mathematically model the stationary random components as belonging to the same distribution type but with no common parameters and determine the appropriate threshold failure probability directly. (See Figure 12.3 where at any given time, a normal distribution is applied, each with a different mean and standard deviation.)

Example: Turbine Blade Cracking

Consider turbine blades tested for crack propagation, where failure is defined as a crack of length 30 mm or greater. Turbine blades would usually be

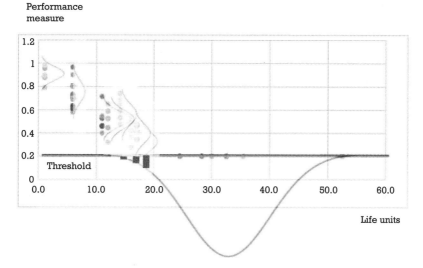

Figure 12.3 Destructive degradation analysis (least preferred option).

cyclically stressed and inspected every (say) 100,000 cycles for crack length and so the nondestructive testing analysis technique described earlier would usually apply. However, consider for our purposes that the records that associate crack lengths with particular turbine blades have been lost (see Table 12.2, from [3]).[3] Therefore, we must consider that each crack length measurement at each 100,000 cycles are from separate turbine blades. Nevertheless, we can still use the 'destructive testing' analysis techniques described to extrapolate data and make conclusions.

We perform a Weibull analysis for the data set at each 100,000 cycles. This shows (regression) shape parameter values for crack size of 4.02, 4.06,

Table 12.2
Turbine Blade Cracking Data
(Crack Length in mm)

Cycles (× 1,000)					
100	15	10	17	12	10
200	20	15	25	16	15
300	22	20	26	17	20
400	26	25	27	20	26
500	29	30	33	26	33

---

3. The table is adapted from [3], which does associate crack length measurements with particular turbine blades.

6.39, 7.84, and 10.16 respectively (with a mean of 5.6) and scale parameter (characteristic life) values for crack size of 14.1, 20.1, 22.5, 26.3, and 31.6 cm respectively.

In the first instance, we might consider applying these characteristic life values to an exponential degradation trend model, with a resulting characteristic life of around 457,500 cycles at the threshold value of 30 cm (and using the average shape parameter value of 5.6 gives a MTBF value of around 423 (x1,000 cycles). However, given that the shape parameter values at each 100,000 cycles are not particularly constant but rather increase significantly, this approach is unsatisfactory.

Instead, we use the alternative approach. We use each Weibull distribution fit at each 100 (x1,000) cycle period to determine the probability of failing (exceeding the threshold value), and then to fit an appropriate distribution to these threshold probabilities. Doing so yields the following probabilities of failure at each cycle time: 0.0000, 0.0064, 0.0017, 0.0585, and 0.5513. Then using a least-squares fit in Solver (see Chaqpter 2 appendix) to fit an associated Weibull distribution of time to fail at the threshold value, we obtain a shape parameter value of 11.6 and scale parameter (characteristic life) value of 509.64 (x1,000), giving an MTBF of around 488 (x1,000) cycles.

This very high Weibull shape parameter value shows rapid onset failure around the MTBF value. Therefore, a failure-free period may be present and a more complex (3-parameter) Weibull model might be considered using more advanced tools.

NOTE: If we had more than five data points to fit a 3-parameter Weibull model, more advanced readers might explore using the Excel Solver tool (see Chapter 2 appendix), for example, by using: WEIBULL.DIST, (IF($(t - \gamma) < 0$,0, $t - \gamma$), β, η, TRUE) where γ is the failure free period. However, in this case with four (or even three or two) data points to fit an equivalent 2-parameter Weibull distribution to (after the failure free period is accounted for), we find that Solver converges to many alternative solutions depending on the starting conditions provided, and so should not be relied upon.

References

[1] Pan, G., Q. Luo, Y. Wang, X. Li, C. Huang, and M. You, "Reliability Evaluation of Long-Life Products Based on Performance Degradation Data," *Proceedings of the 2019 10th International Conference on Information Technology in Medicine and Education (ITME)*, IEEE, January 2020.

[2] Nelson, W., "Analysis of Performance-Degradation Data from Accelerated Tests," *IEEE Transactions on Reliability*, Vol. R-30, No. 2, June 1981.

[3] Reliasoft Corporation reliawiki.com internet site.

CHAPTER 13

Contents

Monte Carlo Simulation

Bayesian Analysis

Data Mining Methods

Prognostics and Health Management

Preview of Advanced Techniques

This book has attempted to introduce readers to practical reliability data analysis through applying reasonably straightforward Excel tools. For more advanced reliability analysis applications and techniques, Excel becomes increasingly cumbersome and the authors recommend dedicated products, such as the Reliasoft suite of applications or MINITAB, noting that many of the principles outlined in this book continue to apply. Advanced analysts sometimes also program their own dedicated applications using specialist statistical tool packages available in products such as MATLAB, R, and Mathematica.

We conclude this book by giving readers a sense of more advanced reliability data analysis techniques that might be considered when problem scopes are too complex to render analysis using the techniques described in this book, but without attempting to be rigorous. Four streams of advanced techniques addressed are: Monte Carlo simulation, Bayesian analysis, data mining methods, and prognostics and health management (PHM).

Monte Carlo Simulation

Monte Carlo simulation is one of the most straightforward simulation techniques. It can be applied when we can characterize base inputs (such as component reliabilities) but the diversity of these inputs make mathematical formulas for outputs (such as system reliability) difficult if not impossible. In this context, the technique relies on (pseudo-)random numbers to generate random times to failure based on a failure distribution. This is applied for all input distributions and combined to produce a sample output value. When repeated thousands (or hundreds of thousands) of times, the technique produces an empirical output distribution. Accordingly, Monte Carlo simulation might be better described as a technique for reliability prediction than reliability analysis, though it can also be used to validate derived analysis models.

Excel contains a (pseudo-)random number generator function RAND(), which generates a uniformly distributed (pseudo-)random number between 0 and 1 (whose value can be updated by pressing the Recalculation Key F9). By using this function, we can generate random numbers having any other distribution, for example:

- Exponential distribution: =−LN(RAND())/λ;
- Weibull distribution: =LN(n*(− LN (RAND())^(1/β));
- Normal distribution: =NORM.INV(RAND()μ, σ);
- Lognormal distribution: =LOGNORM.INV(RAND()μr, σr;
- Uniform distribution: =max_time * RAND().

As an example of how these inputs can be combined, consider a simple two module system where failure of any module results in system failure (a series functional block diagram). In a single Excel row, we model the module failure times using the appropriate functions above in Cell 1 and Cell 2 respectively (T_a and T_b). Then in Cell 3 we enter an appropriate mission time of interest (T_0). In Cell 4, we apply a logical expression to determine if the mission was successful (yielding a 1) using: =IF (AND(($T_a > T_0$),($T_b > T_0$)), 1, 0). Additional rows of these columns represent additional Monte Carlo runs, with the reliability estimated by dividing the number of 1s by the number of runs (e.g., 10,000 runs = 10,000 rows). The Excel data analysis histogram tool can provide output distribution plots.[1]

---

1. When at least one module needs to survive the mission (a parallel reliability block diagram) use: IF (OR((Ta >To),(Tb>To)), 1, 0). Excel add-ins, such as @RISK, make this process more straightforward. Excel macros can also make automating recalculation, varying parameters or result collection simpler.

Bayesian Analysis

The term Bayesian derives from the 18th century mathematician (and theologian) Thomas Bayes, who provided the first mathematical treatment of statistical data analysis using what is now known as Bayesian inference. Bayesian probability seeks to make use of all available relevant (prior) information to evaluate a probability, not just the information from a one-off test or data set. This approach is particularly attractive for reliability analysis given the paucity of data often characterizing the field. Instead of relying on the frequency of some phenomenon to determine objective probability in a traditional sense, the Bayesian approach interprets probability as reasonable expectation of some phenomenon, which can be extended to quantifying our 'subjective' degree of personal belief.[2]

The Bayesian approach first specifies some prior probability or probability distribution based on available relevant information—our reasonable expectation or degree of belief—such as reliability performance of similar equipment or technologies used in roughly similar conditions, or even expert judgement. This prior probability is then updated to a posterior probability in the light of new, relevant data (or evidence). The Bayesian interpretation provides a standard set of procedures and formulas to perform this calculation, based on Bayes Theory (see Chapter 1 appendix). Subsequently, the posterior distribution becomes the next prior. Bayesian methods are now widely used (e.g., in machine learning).

Bayesian methods are, in general, mathematically intensive, with numerical integration or simulation techniques often being required in practice. An exception occurs if the posterior distributions are in the same distribution family as the prior distribution (then termed a conjugate prior); here the expression for the posterior distribution can often not only be calculated in a straightforward way, but the conjugate prior often provides an intuitive sense of Bayesian updating.

For example, the conjugate prior of the binomial distribution is the Beta distribution (associated Excel functions are BETA.DIST and BETA.INV). Therefore, if the Beta distribution can reasonably describe the physical phenomenon under study, we might regard the (hyper)parameters of the Beta distribution α, β) as pseudo-observations or virtual-observations indicating our prior degree of belief: $\alpha - 1$ successes and $\beta - 1$ failures.

The posterior probability distribution then is a Beta distribution with the number of success and failures adjusted according to the new data (e.g.,

---

2. A key philosophical difference with Bayesian inference is that the probability that can be assigned to a hypothesis can be in a range from 0 to 1 if the truth value is uncertain. Frequentists may disagree with the underlying subjectivist philosophy underlying Bayesian approaches since, for frequentists, a hypothesis must be either true or false.

the number of successes and number of failures from a test are simply added to each of our original prior parameter values). The intuition this provides can also help choose reasonable (hyper) parameters for the prior to reflect the relative strength of the prior and the evidence.[3]

NOTE: If the posterior mode (most frequent value) is used to choose an optimal parameter setting, the Beta distribution mode is $(\alpha - 1) / (\alpha + \beta - 2)$ corresponding to $\alpha - 1$ successes and $\beta - 1$ failures. In the absence of data of any kind to inform a prior distribution, noninformative priors appear to be a straightforward solution. The (Jeffreys) noninformative prior distribution for a parameter can be viewed as a uniform distribution for the parameter (corresponding to a maximum entropy), conveniently represented by the Beta distribution with $\alpha = 1$ and $\beta = 1$. For example, a subject matter expert might provide the 5th, mode, and 95th percentiles to determine the parameters of a Beta prior distribution. (Confusing to novices however is that the somewhat more mathematically convenient but less intuitive mean is $\alpha / (\alpha + \beta)$, which corresponds to α successes and β failures.)

Example: Pump Failures Revisited

Chapter 10 considered a success-fail test of a pump, when it failed to start when required in 4 of 240 attempts. The proportion of successes of this pump was calculated as 236/240 = 0.9833. Suppose, based on historical data, that the proportion of successes of the pump is expected to fail 4 times every 4,000 attempts (i.e., prior expected proportion of successes is 0.999). Using Bayes, we might then adjust the estimate of pump proportion of successes to be (3,996+236)/(4,000+240) = 0.998. Note that in this case the prior estimate essentially overwhelms the data, with only a slight movement away from the prior estimate towards the sample proportion. Note also that we would get a different result if we used as a prior estimate 1 fail every 1,000 attempts (or 0.24 failures every 240 attempts), giving the same prior proportion of successes, but representing less confidence in the prior.

As always though, knowledge of particular failure or degradation mechanisms is important in analysis. Changes to any of the triplet of material, environment, and use (even subtle ones) for an item can sometimes have a profound effect.

---

3. Similarly, the conjugate prior of the exponential distribution is the Gamma distribution, whose parameters can be interpreted as the pseudo number of failures in a life test of pseudo time units, represented by the following Excel function to provide a failure rate:

=GAMMAINV((upper_one_sided_confidence_level), number_of_failures, (1/total_test_time)).

Data Mining Methods

The ideal data to use to predict an item's reliability is the field reliability data for that item in the same operating environment, particularly if we know (for each unit) the initial operation time and failure time, and life history or operating profile. Perhaps the next best data to use is field data based on similar equipment and technologies in similar operating environments. Similarity analysis can also refer to recognizing and using certain patterns in data, to estimate item health. As well as statistical analysis (including Bayesian approaches), data mining can also be used to identify patterns in data.

Both statistics and data mining are concerned with data analysis. Both seek to predict the future outcome or to resolve issues. However, statistics first makes and validates theories to test the data, whereas data mining explores the data first, seeking to detect patterns without demanding any supporting theory. Data mining often comes to the fore with large, multidimensional, and multiformat data sets that are increasingly common nowadays.

Supervised classification is one of the oldest and the most important methods of data mining. In particular, support vector machines (SVM) and artificial neural network (ANN) models have demonstrated high practical utility as classification techniques to categorize data into different classes or to predict future trends in the data.

An SVM learns the optimal separating surface boundary (hyperplane)— the hyperplane which has largest separation margin—from two different classes of input samples after transforming data into a high-dimensional space (where each data point is viewed as a dimensional vector), and then performs analysis of new input samples. SVM is considered as one of the most robust and efficient methods among all well-known algorithms for classification. Their major drawback is their computational cost (low speed in training and test phases) as problems become increasingly complex.

ANN are machine learning models formed by a number of processing element (PE) units arranged in a number of layers to form a network of interconnected neurons. During learning, ANNs iteratively tune weights assigned to data and associated simple nonlinear functions in PEs until they are closer to the training data output. If the learning algorithm is good, ANNs can enable active monitoring of a system and in-situ detection of degradation, as well as reliability of complex systems.

Prognostics and Health Management

Traditionally, to improve safety and reliability, critical subsystems and components were designed or engineered to withstand substantial stresses and redundancy was used where this was not enough. Another more recent approach that has become viable is to monitor the condition (or health) of a system, subsystem or component in situ in real-time (or near real-time), and conduct condition-based maintenance when failure is imminent but has not yet occurred. This approach can often also reduce total cost of ownership by minimizing very conservative safety factors in design (over-engineering), redundancy, and unnecessary maintenance [1].[4]

For example, bearings are found in most mechanical systems with rotational components, and bearings tend to exhibit increasingly larger vibration as they degrade. Therefore, bearing condition (or health) can be assessed through sensors continuously collecting bearing vibration data and when the amplitude of a bearing's vibration exceeds a threshold (set close to the failure point while also allowing time to schedule maintenance), the bearing can be replaced before failure. The viability of widespread real-time condition monitoring and associated condition-based maintenance is the basis of PHM. PHM has become viable due to major advances in information and sensor technologies.

PHM is an enabling discipline of technologies and methods to predict remaining useful life so as to provide advance warnings of failure or enable effective maintenance to extend life [2].[5] Unlike conventional reliability analysis, which mostly provides population-based assessments, particularly for nonrepairable items, PHM allows individualized prediction of remaining useful life.

Advances in information (and communications) technologies has also facilitated other effects. To meet increasing functionality and quality, modern systems are often extremely complex with intricate interactions among subsystems/components and with deeply integrated electronics, making failure detection and diagnosis more difficult. Accordingly, as well as failure prediction, PHM also addresses detection and diagnosis of failures.

NOTE: PHM solutions implicitly cover a complete condition-based maintenance process, originally conceived by ISO 13374 as: acquiring data, manipulating data, detecting state, assessing health, prognosis, and generating advisories. Closely connected core issues are condition detection, diagnosis, and prognosis. Condition detection or monitoring distinguishes anomalous behaviors (including failures) through comparing data against a baseline or reference. One of the detection objectives is to minimize false positives and false negatives. Diagnosis

---

4. Niculita, Nwora, and Skaf [1] reported a reduction between 5% and 15% in maintenance costs of condition based maintenance compared to conventional maintenance relating to a dual fuel engine of liquefied natural gas (LNG) vessels.
5. Prognostics is the process of predicting the future reliability of a product by assessing the extent of deviation or degradation of the product from its expected normal operating conditions; health management is the process of real time measuring, recording, and monitoring the extent of deviation and degradation from normal operation condition [2].

determines the nature of the anomaly through isolating the component or failure mode responsible for the anomaly, and also identifying the cause(s) and extent of the anomaly (or potential failure) or failure. Prognosis might provide a probability of failure or remaining useful life based on a predetermined failure threshold [3]. See also [1].

PHM Design

The pervasiveness of information and sensor technologies also presents challenges relating to potentially drowning in data. We need to decide what data to collect, and how often, before we consider what PHM systems will meet our needs. Therefore, PHM itself is a design process. For example, to be targeted and managed, PHM design for an asset might include a FMECA to decide which components or subsystems are critical to operations and cannot be allowed to run to fail or be subject to simpler time-based preventive maintenance (and which components can [3]).

Then, the type and number of sensors can be identified to meet the most critical needs, and a total cost of ownership assessment made regarding particular PHM options.

NOTE: PHM without such structured analysis becomes not much more than unjustified technology insertion, potentially leading to wasted effort and unnecessary expenditure of resources. Condition-based maintenance enabled by PHM, is not always the most cost-effective type of maintenance. When failures are not critical, we can allow run to fail policies. When the lives can be estimated precisely, scheduled time-based maintenance may be the most cost-effective type of maintenance. Outside these situations PHM is often considered to be the maintenance paradigm with the highest value, through avoiding critical failures, and maximizing useful life while providing appropriate lead-time for logistics management. However, PHM often has higher initial cost and a higher requirement for the field workers compared to traditional fault diagnosis and maintenance programs [3].

PHM challenges

Particularly challenging for PHM of information- and electronics-rich systems is the lack of understanding of the interactions of performance parameters and application environments and their effect on system degradation and failure. For example, many predictive methods are not practically acceptable because many information-rich systems have intermittent failures and no fault found (NFF) rates of 40%–85% [4].[6] Consequently, technologies and methods that account for soft faults (faults that manifest themselves

---

6. Williams et al. reported that NFF failures account for more than 85% of all field failures and 90% of overall maintenance costs in avionics[4].

at the system level without any components in the system being damaged or failing) and intermittent failures are needed.

Further, PHM of electronic systems such as radar presents challenges traditionally viewed as either insurmountable or otherwise not worth the cost. For example, direct measurements of device and circuit level parameters are often not feasible in RF systems. Analog circuits operating above, say, 1 GHz are sensitive to small changes in device parameters, resulting in nondestructive, or operational, failure modes which cannot be easily traced (isolated) to individual components. Inserting remote sensors in or near RF components potentially compromises system performance by introducing noise while affecting overall cost and reliability. Instead, less direct system-level parameters may need to be utilized to identify key prognostic features that correlate with failure progression. With examples of good, bad, and unknown feature sets, classifiers can be developed using an array of techniques from straightforward statistical methods to methods, such as ANN [5].

Other PHM challenges are to develop techniques to combine information from disparate sources to provide optimal fused predictions, and to better quantify uncertainty (e.g., in making maintenance and logistics decisions, predictions in the form of a PDF will be more informative than using point estimates) [6].

Data-Driven PHM

Two general PHM methods are model-based and data-driven methods, each with particular advantages and disadvantages [6]. Since this book is about data analysis, we will focus on data-driven methods.

Data-driven PHM techniques detect system health trends and anomalies by assuming that the statistical characteristics of the system data remain relatively unchanged, and so use statistical patterns and machine-learning to detect changes in parameter data. The most important techniques are time series analysis and stochastic processes that this book addresses, proportional intensity models (see Chapter 2 appendix), as well as more advanced techniques such as Bayesian methods, SVM, ANN, and other methods (such as Markov chains, briefly mentioned in Chapter 11).

One of the advantages of data-driven approaches is that they do not require system-specific knowledge but can learn the behavior of a system based on monitored data (i.e., can be used as black-box models). Further, data-driven approaches can be applied to complex systems where a large number of parameters are monitored. Data-driven approaches have also proven to be suitable for diagnostic purposes. Further, it is possible to detect sudden changes in system parameters allowing for detection and analysis of intermittent faults and reducing no-fault-founds. However, a limitation of data-driven approaches is the requirement for historical training data to determine correlations, establish patterns, and evaluate data trends leading to failure. Sometimes necessary historical data does not exist.

NOTE: Prediction using model-based approaches to PHM is based on knowledge of the fundamental physical processes causing degradation and leading to failure. Modeling these processes allows calculation of damage accumulation and remaining useful life for known failure mechanisms. An advantage of model-based approaches is that, because they take into account degradation caused by environmental conditions, they can be used to estimate damage in situations where systems are in a variety of environments and states. Limitations of model-based approaches include that developing such models requires detailed knowledge of not only the underlying physical failure(s) but also system-specific knowledge. Further, sudden changes in system parameters that characterize intermittent faults are not accounted for in these models. Combining model-based and data-driven methodologies can potentially combine the strengths of each. For example, understanding the physical processes of the critical failure mechanisms can help in choosing the data-driven techniques for diagnosis and prognosis [6].

References

[1] Niculita, O., O. Nwora, and Z. Skaf, "Towards Design of Prognostics and Health Management Solutions for Maritime Assets," *5th International Conference on Through-life Engineering Services* (TESConf), 2016.

[2] Tsui, K. L., N. Nan Chen, O. Zhou, Y. Hai, and W. Wang, "Prognostics and Health Management: A Review on Data Driven Approaches," *Mathematical Problems in Engineering*, Article ID 793161, Hindawi Publishing Corporation, 2015.

[3] López, A. J. G., A. C. Márquez, M. Macchi, and J. F. G. Fernández, "Prognostics and Health Management in Advanced Maintenance Systems," *Advanced Maintenance Modeling for Asset Management: Techniques and Methods for Complex Industrial Systems*, Cham, Switzerland: Springer International Publishing, 2018.

[4] Williams, R., J. Banner, I. Knowles, M. Dube, M. Natishan, and M. Pecht, "An Investigation of 'Cannot Duplicate' Failures," *Quality and Reliability Engineering International*, Vol. 14, No. 5, 1998, pp. 331–337.

[5] Brown, D. W., P. W. Kalgren, C. S. Byington, and M. J. Roemer, "Electronic Prognostics – A Case Study Using Global Positioning System (GPS)," *Microelectronics Reliability*, Vol. 47, 2007, pp. 1874–1881.

[6] Pecht, M., and R. Jaai, "A Prognostics and Health Management Roadmap for Information and Electronics-Rich Systems," *Microelectronics Reliability*, Vol. 50, 2010, pp. 317–323.

Bibliography

Lawless, J. F., *Statistical Models and Methods for Lifetime Data*, John Wiley & Sons, New Jersey, 2003.

Modarres, M., Kaminskiy M., and Kritsov V., *Reliability Engineering and Risk Analysis–A Practical Guide, Second Edition*, CRC Press, 2010.

Nelson, W. B., *Applied Life Data Analysis*, John Wiley & Sons, New Jersey, 2004.

NIST/SEMATECH e-Handbook of Statistical Methods, http://www.itl.nist.gov/div898/handbook.

O'Conner, P., and A. Kleyner, *Practical Reliability Engineering*, Fifth Edition, John Wiley & Sons, 2011.

Reliasoft Corporation internet site www.reliawiki.com.

About the Authors

Darcy Brooker is an engineering and technology manager in the defense sector. Darcy has deep skills and experience in capability acquisition and sustainment roles, such as integration and program director, capability manager representative, project director and manager, and engineer, including seven years as a senior reliability engineer. His engineering and technology knowledge and skills are supported by extensive academic qualifications including master's degrees in IT and computer science, operations research, and engineering. Darcy holds a Master of Engineering degree in Reliability Engineering from the University of Maryland (United States), is a Professional Engineer (CPEng) and Fellow of the Institute of Engineers Australia (FIEAUST), and Certified Reliability Engineer (ASQ).

Mark Gerrand is a reliability engineer and logistics support analyst with extensive reliability and logistics engineering experience in the Aerospace and Maritime environments both within the Australian Department of Defence and the private sector, and across the entire capability systems life cycle. Mark's experience in engineering and logistics includes twelve years as a reliability engineer, seven years as a logistics engineer and logistics support analyst, six years as a fleet systems performance analyst, and three years as a technical author. Mark's qualifications include a Master of Maintenance and Reliability Engineering from Federation University Australia and a Bachelor of Mechanical Engineering (First Class Honors) from the University of Technology Sydney.

Index

A
accelerated life models, 93
accelerated life testing, 175
acceptance (pass/fail) testing, 151, 159
achieved availability, 32
adaptive cost policies, 63
age trend, 38, 41
analysis of variance (ANOVA), 130
analyzing degradation data, 171
Anderson-Darling Test, 62
ANS/ASQ Z1.4, 160
Arrhenius model, 93
artificial neural network (ANN), 183, 186
availability, 23, 32

B
bathtub curve for nonrepairable items, 83
bathtub curve for repairable items, 52
Bayes' Theorem, 36
Bayesian, 36, 63, 88, 181, 186
bell curve. See normal distribution
Benard's approximation, 108
Bernoulli trials, 152
beta distribution, 88, 154, 181
better than old, but worse than new, 41
binomial distribution, 151, 165
built-in test (BIT), 92
burn-in, 41, 83

C
cables, connectors, and solder joints, 92
censored data, 14, 27, 105
Central Limit Theorem (CLT), 69, 130, 133, 139
Centroid Test. See Laplace Test
characteristic life. See scale parameter
Chi-square (c2) goodness-of-fit test, 95
Chi-squared (c2) distribution, 87, 122
Clopper-Pearson principle, 154
combinations, 158
combining event probabilities, 33, 34
common cause failures, 54
Complement Rule, 33
concave curves in Weibull analysis, 113
conditional probability, 35
confidence interval, 25, 68
confidence interval for the normal distribution mean, 133
confidence interval for the normal distribution standard deviation, 134
confidence level, 26
conjugate prior, 181, 182
consumer's risk, 159
contingency tables, 34
continuous random variables, 28
corners, doglegs, and s-curves in Weibull analysis, 114

cost models of repairable items, 62
Cox and Lewis log-linear model, 44
Cox's model, 55
Cramer-von Mises statistic, 59
Crow bounds, 59
Crow-AMSAA model, 44
Crow-AMSAA Test, 60
cumulative distribution function (CDF), 41, 80
cumulative plots, 70

D

data mining, 183
data-driven prognostics and health management (PHM), 186
degrees of freedom, 89
dependability, 23
design of experiments, 130
destructive degradation sampling using a common distribution parameter, 174
destructive degradation sampling with no common distribution parameter, 176
deterioration, 39, 43, 44, 52
discrete random variables, 28
Distribution-EZ, 70, 87, 102, 121, 131, 145
distribution-free. See nonparametric
double sampling, 160
Duane model, 44

E

e, 56
electromagnetic interference/ compatibility (EMI/EMC), 91
empirical cumulative distribution function. See cumulative plots
Excel versions, 12
expected value. See mean
exponential model, 94
exponential distribution, 41, 119
exponential functions, 56
extreme value distributions, 88
Eyring model, 94

F

F distribution, 88, 154
factorial, 158
failure modes, 38, 89
Failure Modes Effects Analysis (FMEA), 55
Failure Modes, Effects, and Criticality Analysis (FMECA), 12, 185
failure-censored data, 28
failure-free period, 101
fault tree analysis (FTA), 12
Fisher matrix, 69
FMECA, See Failure Modes, Effects, and Criticality Analysis (FMECA)

G

Gamma distribution, 88, 182
Gaussian distribution. See normal distribution
General Additive Rule, 34
General Multiplication Rule, 35
gentle curves in Weibull analysis, 113
goodness-of-fit, 37, 57, 59
goodness-of-fit tests, 95, 96
GRP-EZ, 46

H

hazard rate, 52, 82, 119
hidden failures, 33
histogram, 69
homogeneous Poisson process (HPP), 42, 43, 53, 163
HPP, See homogeneous Poisson process (HPP)
human error, 20, 53
human factors, 31
human reliability analysis, 19
hypergeometric distribution, 88, 152

I

identically and independently distributed (IID), 37, 39, 41, 42, 44, 50, 51
IID, See identically and independently distributed (IID)

Index

independence, 29, 42
independent events, 36
infant mortality, 83
inherent availability, 32
integrated circuits and nonmicroelectronic component reliability, 91
interarrival times, 43
intermittent failures, 54, 92, 185
interval censored data, 71
inverse power law model, 94
ISO 2859-1(1999), 160

J
Jeffreys prior, 182
joint and marginal distributions, 82

K
Kaplan-Meier method, 71, 75, 109
Kijima virtual age models, 46
Kolmogorov-Smirnov (K-S) goodness-of-fit test, 96

L
lag, 55
Laplace Test, 60
least squares curve fitting, 57
least squares estimate, 48
life units, 22, 54
limit theorems, 42, 53
lognormal distribution, 141
lognormal distribution parameter confidence limits, 145
lognormal distribution parameters, 144

M
maintainability, 23, 31
maintenance, 31
Mann Test, 61
Mann-Whitney U test, 73
margin of error (E), 133
Markov chains, 186
Markov models, 167
maximum likelihood estimate (MLE), 45, 106
mean, 66

mean and standard deviation of raw lognormal data, 146
mean residual life, 85
median, 66
median rank, 70
median test, 73
memoryless, 119
MIL-HDBK-189 Test, 62
MIL-HDBK-217F, 90
MIL-STD-105, 160
MIL-STD-1916 (1996), 160
minimal repair, 40
mode, 66
Monte Carlo simulation, 167, 180
MTBF, 32, 44, 67, 121
MTTF, 67, 121
MTTR, 67
mutually exclusive events, 34

N
no-failure confidence, 69
nondestructive degradation sampling, 172
nonhomogeneous Poisson process (NHPP), 40
nonparametric analysis, 38, 65
nonparametric tests, 72
nonrepairable items, 38
normal distribution, 129, 165

O
one-shot devices, 39, 156
operational availability, 33
ordinary renewal process, 40

P
parameter, 25
parts count methods, 90
PDF, See probability density function (PDF)
point and interval estimates for binomial distribution parameter, 152
point and interval estimates for Poisson distribution parameter, 168

point and interval estimates for the exponential distribution parameter, 121
point estimates, 25
point estimates for the normal distribution parameters, 132
Poisson distribution, 163
Poisson process, 119
population, 24
posterior probability. See Bayes' Theorem
power function, 43
power law model, 44, 58
Prentice-Williams-Peterson (PWP) model, 55
preventive maintenance modeling, 46
prior probability. See Bayes' Theorem
probability, 11, 24
probability density function (PDF), 41, 81
producer's risk, 159
prognostics and health management (PHM), 171, 183
proportional hazard models, See proportional intensity models
proportional intensity models, 38, 44, 46, 54, 186

R
R2 coefficent of determination, 57, 105
RAM, 21
random, 24, 27
range, 67
rank adjustment method, 108
rare event approximation, 34
rate of change of failures (ROCOF), 32, 40, 52, 120
rate of occurrence of failure, See rate of change of failures (ROCOF)
Rayleigh distribution, 101
Reactor Safety Study, 53
reliability, 22
reliability block diagram (RBD), 12
reliability growth, 39, 43, 52

renewal, 40, 42, 46, 50, 53
repairable item failure data, 41
repairable items, 38, 50
repairable system growth, 44
residuals, 175
right-censored data, See censored data
risk, 26
ROCOF, See rate of change of failures (ROCOF)

S
safety, 23
safety analysis, 14
safety margin, 84
same as old, 40
sample, 24
sample statistics, 66
scale parameter, 100
scatter graph, 43
sequential sampling, 160
shape parameter, 100
sign test, 73
smallest extreme value distribution, 99
software reliability analysis, 19
spares assessing, 166
standard deviation, 67
standardized normal variate (z) 136
stationary, 40, 42
statistic, 11, 25
statistical process control (SPC), 130
statistical significance, 25
stochastic point process, 39, 51
Student's t distribution, 88
Sturges' Rule, 69
support vector machines (SVM), 183, 186

T
testing for independence, 55
tests for linearity, 60
thermal design, 91
time censored data, 28
transient voltage protection, 91
trend analysis, 43
trends, 39, 41

Index

U
unbiased, 65
useful life, 84

V
variance, 67
virtual age models, 46

W
waiting line paradox, 164

wear out, 43, 52, 84
Weibull distribution, 44, 99
Weibull probability plotting, 105
Wei-EZ, 107
Wilcoxon rank sum test, 73
Wilcoxon signed rank test, 73

Z
zero failures, 69, 154

Recent Titles in the Artech House Technology Management and Professional Development Library

Bruce Elbert, Series Editor

Actionable Strategies Through Integrated Performance, Process, Project, and Risk Management, Stephen S. Bonham

Advanced Systems Thinking, Engineering, and Management, Derek K. Hitchins

Critical Chain Project Management, Third Edition, Lawrence P. Leach

Decision Making for Technology Executives: Using Multiple Perspectives to Improve Performance, Harold A. Linstone

Designing the Networked Enterprise, Igor Hawryszkiewycz

Electrical Product Compliance and Safety Engineering, Steli Loznen, Constantin Bolintineanu, and Jan Swart

Engineering and Technology Management Tools and Applications, B. S. Dhillon

The Entrepreneurial Engineer: Starting Your Own High-Tech Company, R. Wayne Fields

Evaluation of R&D Processes: Effectiveness Through Measurements, Lynn W. Ellis

From Engineer to Manager: Mastering the Transition, Second Edition, B. Michael Aucoin

How to Become an IT Architect, Cristian Bojinca

Introduction to Information-Based High-Tech Services, Eric Viardot

Introduction to Innovation and Technology Transfer, Ian Cooke and Paul Mayes

ISO 9001:2000 Quality Management System Design, Jay Schlickman

Managing Complex Technical Projects: A Systems Engineering Approach, R. Ian Faulconbridge and Michael J. Ryan

Managing Engineers and Technical Employees: How to Attract, Motivate, and Retain Excellent People, Douglas M. Soat

Managing Successful High-Tech Product Introduction, Brian P. Senese

Managing Virtual Teams: Practical Techniques for High-Technology Project Managers, Martha Haywood

Mastering Technical Sales: The Sales Engineer's Handbook, Third Edition, John Care and Aron Bohlig

The New High-Tech Manager: Six Rules for Success in Changing Times, Kenneth Durham and Bruce Kennedy

The Parameter Space Investigation Method Toolkit, Roman Statnikov and Alexander Statnikov

Planning and Design for High-Tech Web-Based Training, David E. Stone and Constance L. Koskinen

A Practical Guide to Managing Information Security, Steve Purser

Practical Model-Based Systems Engineering, Jose L. Fernandez and Carlos Hernandez

Practical Reliability Data Analysis for Non-Reliability Engineers, Darcy Brooker with Mark Gerrand

The Project Management Communications Toolkit, Second Edition, Carl Pritchard

Preparing and Delivering Effective Technical Presentations, Second Edition, David Adamy

Reengineering Yourself and Your Company: From Engineer to Manager to Leader, Howard Eisner

The Requirements Engineering Handbook, Ralph R. Young

Running the Successful Hi-Tech Project Office, Eduardo Miranda

Successful Marketing Strategy for High-Tech Firms, Second Edition, Eric Viardot

Successful Proposal Strategies for Small Businesses: Using Knowledge Management to Win Government, Private Sector, and International Contracts, Sixth Edition, Robert S. Frey

Systems Approach to Engineering Design, Peter H. Sydenham

Systems Engineering Principles and Practice, H. Robert Westerman

Systems Reliability and Failure Prevention, Herbert Hecht

Team Development for High-Tech Project Managers, James Williams

For further information on these and other Artech House titles, including previously considered out-of-print books now available through our In-Print-Forever® (IPF®) program, contact:

Artech House
685 Canton Street
Norwood, MA 02062
Phone: 781-769-9750
Fax: 781-769-6334
e-mail: artech@artechhouse.com

Artech House
16 Sussex Street
London SW1V 4RW UK
Phone: +44 (0)20 7596-8750
Fax: +44 (0)20 7630-0166
e-mail: artech-uk@artechhouse.com

Find us on the World Wide Web at: www.artechhouse.com